BLACK LION

BLACK LION

Alive in the Wilderness

SICELO MBATHA AND
BRIDGET PITT

Jonathan Ball Publishers
Johannesburg · Cape Town · London

© Text Sicelo Mbatha and Bridget Pitt 2021
© Published edition 2021 Jonathan Ball Publishers

Originally published in South Africa in 2021 by
JONATHAN BALL PUBLISHERS
A division of Media24 (Pty) Ltd
PO Box 33977, Jeppestown 2043

This edition published in 2021 by Jonathan Ball Publishers
An imprint of Icon Books Ltd, Omnibus Business Centre,
39-41 North Road, London N7 9DP
Email: info@iconbooks.com

For details of all international distributors, visit iconbooks.com/trade

ISBN: 978-1-77619-148-2
eBook ISBN: 978-1-77619-129-1

Cover by Luke Bird
Cover illustration by Afro Deco
Photos: p. 20, top and bottom: Bridget Pitt; p. 39: Bridget Pitt; p. 51: Bridget
Pitt; p. 56: Christopher Bartlett; p. 78: Bridget Pitt; p. 118: Wayne Saunders; p.
126: Geseko von Lüpke; p. 131: Bridget Pitt; p. 143: Wayne Saunders; p. 160:
Wayne Saunders; p. 174: Thabani Nzuza; p. 186: Siphesihle Ngcobo; p. 189:
Anya Mendel; p. 193: Bridget Pitt; p. 204: Bridget Pitt; p. 213: Richard
Knight; p. 214: Bridget Pitt; p. 215, top and bottom: Richard Knight; p. 224:
Adam Altenburg; p. 234: Richard Knight; p. 243: Bridget Pitt; p. 246: Cebo
Zikhali; p. 248: Sicelo Mbatha; p. 250, left and right: Bridget Pitt; p. 251:
photographer unknown; p. 255: Gregory Buchhaus; p. 260: Andiswa Mbatha;
p. 261: Simone Henke; p. 263: Simone Henke; p. 264: Bridget Pitt; p. 269:
Sicelo Mbatha; p. 274: Geseko von Lüpke

Printed and bound in Great Britain
by Clays Ltd, Elcograf S.p.A.

CONTENTS

———

PROLOGUE

———•———

WE CAME ACROSS HIM SUDDENLY. AS WE EMERGED FROM THE REEDS, there he was on the riverbank: a fully grown male lion, hardly three metres away. His pitch-black mane was matted with the blood of his last kill; flies hovered over his sinewy body, their buzzing thick in the afternoon heat.

For a moment I thought he was dead, because of the pungent scent of blood and the flies. But his position told me that he was sleeping. As I looked more closely I saw that his eyes were twitching, his nostrils expanding and contracting with each breath. My blood ran cold. He was very much alive, and just one paw slash away. And I was all that stood between him and the eight trailists in my care.

I swore softly under my breath as I loaded my .458 Brno. At the click of the bolt as it took the round into the chamber, his eyes flew open. He leapt up, growling, his blazing eyes scorching mine. I sensed my backup guide retreating slowly to a safe place with the trailists, heard the growls and rustling of another lion emerging from the reeds behind us and running across the river. The black-maned lion stood before me, snarling and twitching his tail, and slapping at the ground with his massive front paws. The thunderous rumble of his growling reverberated through my stomach, turning my insides to liquid. My knees were shaking so hard I could barely stand, but I did, my firearm pointed at him, my finger resting on the trigger, ready to shoot.

I knew any mistake could kill me, that he was ten times stronger than I was, that even if I fired the gun as he leapt, the shot might go wide or the gun jam. But I also knew that, from the depths of my heart, I did not want to shoot him. My deepest intention was to leave him unharmed, and my best chance of walking away unharmed myself was to communicate this intention. And so I spoke, softly, more in my mind than aloud.

'Calm down, my brother. I know you are powerful. I did not mean to disturb you. Let us go on our way and we will leave you in peace …'

As I spoke, the intention behind the words seemed to reach him, and gradually he began to calm down, while keeping his gaze on me. But when I took a step backwards, he growled again and slapped the earth, and his hot pungent breath was thick in my nostrils. I kept inching backwards, kept talking to him, kept my eyes on his. Slowly his growls softened to a rumble. Slowly, his mane lowered. He dropped his shoulders and sank to the ground. A deep silence overcame everything; no birds were calling, the wind was still, the baboons that had been barking up on the hills had gone quiet.

When I had backed a good distance away, I turned around and walked back to the group, who were sitting under a splendid thorn some seventy metres from the lion. I gestured to my backup guide – a young woman whom I was mentoring called Bonangiphiwe Mbanjwa, or Bona – to keep watch, then lay down on a bed of leaves, absorbing what had happened. My body trembled with the power of this encounter: the memory of his burning eyes, the reverberations of his growling. I felt my soul leap within me, enlivened by the lion's courage and wisdom, by the sense that I'd been touched by something mysterious and divine. My body sang with the knowledge that the lion was sleeping seventy metres away. My heart quietened as I listened to the sweet piercing song of the gorgeous bushshrike from the depths of a nearby magic guarri bush, until the sound carried me into sleep.

It is this encounter that gave me the wilderness name Black Lion. One of the trailists called me this, remarking on how the lion and I had mirrored each other – the lion with his heavy black mane, me with the abundant black beard that I had at that time. I embraced this name not out of arrogance nor believing I had 'faced down' a lion. I took it on as a reminder always to embrace the wild lion inside me; to instil in myself the power of sharing, as the lions share their kills with the pride, and with the hyenas and vultures that follow; to instil in myself the power to fiercely protect what I love, as the lioness protects her cubs; to instil humility and deep respect for the creatures who share this earth; and to remind myself that life is precarious, each moment a gift to be savoured.

I am not a lion whisperer. I am an ordinary wilderness guide who has simply grasped some small part of the great wisdom of wild animals and the wilderness, and learnt a little about how to share this with other humans. As a guide, I share wild paths with wild creatures; I swim in the waters of wild rivers, where animals drink and wallow with freedom. I embrace both the scorching sun and the tumultuous midnight thunderstorms.

My wilderness stories do not come from me. They have been told by rivers and streams, by the lonely buffalo and the woodland dove, by fluttering butterflies and singing frogs and the slow creeping chameleon, by fragile flowers blooming in a decaying stump in the heart of the wilderness. These elements have birthed me, and I am one with them. I breathe the same air as the lion roaring for the moonrise, as the dung beetle foraging underfoot.

Walking in the wilderness has enriched my life beyond the telling of it. Every encounter with wild creatures has brought symbolic messages to me, teachings that I could never have found in a textbook. It has been my life's path to rekindle the wildness in all of us, to bring people into the presence of wildness and help open their souls to its beauty, wisdom and infinite power to heal.

I am the black lion who helps people discover the wild animal within.

I am the black lion who roars for peace and harmony on Great Mother Earth.

I am the black lion, alive in the wilderness.

Part One

———•———

THE CROCODILE'S GIFT

MY NAME IS SICELO CABANGANI MBATHA. MY WILDERNESS NAME IS Bhubeselimnyama – Black Lion. I am the youngest son of Mtukayise Thunduzela Mbatha, who is the one and only child of Phondolwendlovu Mbatha. My grandfather had three wives – my grandmother died when she gave birth to my father. I come from the Zulu tribe under the Mbatha clan.

The Mbatha are Nguni people who originated from Mageba, the son of Zulu, through his grandson Sontshikazi. Mageba was twin brother to Phunga, who became a chief of the Zulu clan. Phunga and Mageba were separated during infancy because they were twins and it was taboo to the Nguni people of the time for twins to grow up in the same household. Mageba grew up at his maternal home at Nkandla so that he wouldn't be a threat to Phunga, who was the heir to the Zulu throne.

I was born and raised in the deeply rural area of kwaHlabisa, on the doorstep of what was then called the Hluhluwe Game Reserve, now the Hluhluwe–iMfolozi Park. I grew up steeped in nature. From my earliest days I was thirsty for all the wisdom that the wild plants and animals, the rolling hills and wide skies, could bring to me. I have been lucky to receive teachings from more living beings that I can count – from the fragile butterfly to the great, ponderous elephant. But before we wander further down the winding paths of my life's journey, I would like to tell you the story of the crocodiles' cruel but healing gift to me, for this story shows us just how deeply and profoundly nature can bring healing and wisdom to the troubled soul.

It was always my ambition to study nature conservation, but my parents lacked the means to pay for my studies. I therefore decided to work as a volunteer at the Hluhluwe–iMfolozi Park. On a winter's day, I was walking with Baba Thabethe and Dumisane Khumalo on patrol along the banks of the great iMfolozi River. We walked along a river parched by the dry season – a snaking ribbon of white sand, scattered with a few small pools, packed with water insects and desperate tilapias. Hovering kingfishers swooped down, feasting on the trapped fish and insects.

The winter wind spun into a whirlwind, sending dry leaves swirling into the sky. I sneezed as it enveloped us and the dust penetrated my nostrils. My colleagues laughed when I tried to escape it, calling, 'Sokugwinya isikhwishikazane' – the whirlwind will swallow you. A zebra stallion shied away in alarm, before galloping off with his family through the leafless bushes, their hooves pounding the earth.

A lonely waterbuck stood silently watching us as we scanned the reeds to assess whether it was safe to enter – patrols through the reeds can be lethal, as predators are camouflaged and it is easy to stumble upon them.

The warning calls of vervet monkeys and banded mongoose alerted us to a possible threat deep in the reeds. The chorus swelled as birds joined the cacophony. We entered and, walking in single file, cautiously approached the source of the noise. As we drew closer the sounds died down, for now the birds and mongooses were frightened by our presence. Soon, the only sound was the soft sighing of the reeds, the suck and slosh of mud, and the laboured breathing of a heavy animal.

Then we heard a deep hissing, and jaws snapping against a soft body. As we came through the reeds, a shocking sight confronted us. A big male buffalo was sunk up to his belly in a mud pool, while crocodiles feasted on his flesh. He was still alive, but helpless.

It was a horrifying scene. But for me, a greater horror lay in the images that exploded from my memory, to match the terror and agony of the buffalo before me. As the buffalo's blood filled the mud hole, turning the grey water to red, one word filled my mind.

Sanele.

Sanele was my godfather's child, two years older than me – a bright, lively boy, always laughing, my soul mate and my hero. Sanele knew everything about tracking birds and animals, about where to find the sweetest wild summer fruits, the bushman plums and sour plums and water berries, which sustained us on our daily fourteen-kilometre journey to school and fourteen back. A childhood spent herding goats and running wild in the veld had made us all lithe, fit and strong. But Sanele was the strongest of all.

Sanele was the best swimmer among us, and we relied heavily on his help to get across the rivers. There were three between my home and school. In winter, they were little more than streams. But the summer storms could quickly turn them into raging torrents. When the teachers saw the rain coming down, they would let the younger children out early to get home before the waters rose. But this act of mercy could also put us in danger, for we needed the older children to help us across.

When the older children weren't there, Sanele was the one to guide us through the water and would rescue our 'plastics' – the plastic supermarket bags that we used to carry our school books – when we lost our grip on them and they floated away.

One December day, just after I turned seven, the rain was coming down hard in our part of Zululand, and the teachers sent us young ones home early as usual. We crossed the first two rivers with some difficulty. The water was flowing fast and up to our waists, much higher than usual, and the rocks were slippery underfoot. We knew that it would be tough to cross the last river, as this one was the deepest.

We stood on its banks, contemplating the muddy water racing past. Some wanted to wait for the older children. But they would not come for some hours, it was raining hard, and we were wet and freezing cold. We tried to find a better place to cross, but our minds were numbed by cold and exhaustion. At length we decided to cross at the usual spot, holding one another's hands in a line.

Before we ventured in, we scrutinised the water for crocodiles. We knew that they might be around, for they occasionally caught a dog or a goat. If you see a log floating upstream, it's a crocodile, the adults warned us.

We could see no logs floating upstream, nor could we see any debris floating down that might knock us off our feet, so we stepped reluctantly into the cold, turbid water. Sanele was walking in front, holding my left hand, then me, then two or three girls. About halfway across, a log knocked against us as it swept past, breaking the line and causing one girl to stumble and drop her plastic. She lunged for it, and fell again, and we told her to leave it.

Just before we reached the far bank, the girl at the back of the line cried, 'Crocodile!' As I turned to see where she was pointing, a powerful jolt came from Sanele's hand, and it was wrenched from my grasp. I swung back to him, but he had disappeared under the water – only his hand was above the surface, clutching at the air. The water was churning, and I could see the crocodile's back, thrashing in the foam. I grabbed Sanele's hand again, and tried to pull him towards the side. One of the girls was also trying to pull him; the other was standing crying on the bank. Numb with terror, I clung to my friend's hand. But it was slipping through my cold, wet fingers. Then I saw a bloom of blood, turning the muddy brown river red. A fountain of blood spurted out of the water, shooting up and spraying my white shirt. I felt Sanele's hand grow limp in mine, as if his spirit had left his body. I knew then that he had lost the battle with the crocodile, yet still I gripped his hand harder, as hard as I could, as hard as if it were my own life that I was clinging to.

But his fingers slipped from my grasp.

Sanele was gone.

We scrambled out of the water, and ran down the bank, hoping that the crocodile would leave him. But he was nowhere. All we could see was his plastic, spinning away with the current. Then that, too, was gone.

The elders came with spears to find the crocodile, but there was no sign. All they ever found was Sanele's T-shirt, two weeks later, caught on a branch downstream.

———————

I had to go to school the next day. I had to cross the same river, in the same place. I was consumed by terror, for I was so sure the crocodile would take another one of us. When I got to school, I had to write a mental arithmetic test to pass into Grade 2. In what world can a child who has lost a friend like this be expected to get up the next day and do mental arithmetic? For a while the adults took it in turns to cross the rivers with us. But the rainy season is the busy season in rural Zululand. Fields must be ploughed; crops must be planted or harvested so that empty stomachs can be filled. Within a few days, their work took them back to the fields, and we were left to cross the rivers as best we could.

No one counselled me. No one cosseted me, or helped me to grieve. I felt like an iron, burnt red-hot in a fire, then hammered into shape and plunged into icy water. Losing Sanele to the crocodile was horrific. But every day that I had to endure without him was worse. I felt as if I'd lost a limb, as if my heart had been torn from my body.

For weeks I was lost in a dark thicket of grief and fear. I could not sleep, I could not eat. How would I ever cross the rivers without Sanele by my side? How could I walk up the long, steep hills, or eat our favourite food of imifino (spinach) leaves and steamed cornbread, or track the guineafowl and eagles, without Sanele by my side? How was even one day of my life imaginable without Sanele by my side? I almost walked into the valley of suicide, so devastated was I by the depression and fear that I endured. The shadow of his death haunted every breath I took.

My grief morphed into anger, then swelled into a cold, black, bottomless pool of hate. How could one small boy hold so much hate? I hated the sun for bringing yet another day without Sanele. I hated school, and every step of the fourteen-kilometre walk there, and every step of the fourteen-kilometre walk back. I hated the rivers. I hated the adults who made me go to school, the children who weren't Sanele. I absolutely, violently and vehemently hated all crocodiles, and swore to avenge my friend's death.

Drowning in the centre of this deep, black pool of hatred was me. For I had failed to save my friend. His life had been in my hand. And I had let it go.

It lived with me for a long time, that hate. As the months and years went by, I learnt to push it aside. I grew a skin over it, but it lay deep within me, a festering splinter of pain. Until the day I saw the crocodiles tearing at the buffalo.

———————

As I stood watching the crocodiles rip the buffalo's intestines, I was seven years old again and back in that river. I could hear Sanele's last strangled cry as the crocodile pulled him under. I could see his frantic face, as he gasped for breath. I could feel his hand go limp. And, again and again, I could feel his cold, wet fingers slipping through mine. I had buried this pain, but here it was before my eyes, sinking its teeth into me as if it were my flesh being torn by the crocodiles. I could not turn away.

We watched the light dying in the buffalo's eyes. His head drooped and his heavy horns sank.

'Now he is at peace,' said Baba Thabethe. He was our patrol leader, an old, strong ranger with much wisdom.

'His old life is ending now,' he continued, 'but his new life is beginning. The buffalo is at peace, but there are many people alive who have no peace. They may seem successful, but inside they are dying, for they cannot make peace with their past lives. We need to think about what this death of the buffalo can mean for us.

'Just as the buffalo had to die before his new life could begin, so we need to understand that sometimes one part of us has to die to allow a new part to grow. You cannot have the old part and the new part living together. The sun and the rain cannot share the sky. For the sun to come, the rain must go.'

His words calmed the whirlwind of horror that my memories had raised. Vultures circled high above us or perched in a nearby umkhiwane (sycamore fig) tree, waiting patiently for us to go. The soft whooping of hyenas in the stream beyond the reeds encouraged the crocodiles to eat more quickly. The buffalo's life had ended, but it was giving life to other beings, and so the circle would continue.

As we made our way back through the reeds, I understood that I needed to let go of the hatred and sorrow and anger I had carried for all those years. I realised that I no longer held hatred for the crocodiles in my heart. For years I had associated crocodiles with fear and brutality. But the crocodile that took Sanele was just taking the opportunity to get a meal – it was not acting out of cruelty or vengeance. Finally, I could accept crocodiles as fellow creatures, even worthy of respect – for they are formidable survivors, able to live without water for several days, to survive without food for months.

I followed the others along the river, my mind churning with thoughts and emotions. The experience seemed to be inviting me to walk across a bridge and leave my past behind. But it also helped me understand why doing this had been so difficult. Walking across the bridge demands a brave heart, brave enough to face what is hurting your soul and to make peace with it. I had not been able to face that pain, for I was just a young boy, and given no time or space to lament, no time to grieve. These crocodiles were giving me a chance to be free, just as they were freeing the buffalo that was stuck in the mud. For however much he suffered while they were eating him, if left he would have suffered the much slower death of starvation.

I could not let my pain go until I had acknowledged it, its depth and breadth, and had felt, again, its sharpness. The crocodiles had given me this cruel and beautiful gift – perhaps they alone could bring this message to me. Just as the crocodile had taken Sanele, so it was the crocodiles that restored me to myself. Guided by the wisdom of Baba Thabethe, I understood that the wilderness was enabling me to face what had been eating my soul for the past fifteen years. It came as bitter medicine, but I knew it would heal when swallowed.

I looked at the dry riverbed, scattered with shrinking pools. The animals were suffering in those pools, but it felt as if the river was preparing itself for the new waters of the spring, for the new life that these waters would bring. I imagined the cool, sweet waters rushing over the white sand, and felt forgiveness flowing over my soul. Forgiveness for the crocodiles – but, more importantly, forgiveness for myself. Forgiveness for not having been able to save my friend, and to heal, to create room for new wisdom, until now.

It has taken me many years to heal from Sanele's death, to accept that he was a light spirit, come to earth for a few short years to spread his wisdom before moving on. But my healing began that day. Without the crocodiles' help, I might have remained trapped in anger, hatred and grief, eaten alive as the buffalo was, as I relived the violence that had been done to me, day after day, for a lifetime.

I needed courage to revisit the day I had lost my soul brother, to face the darkness in my past and crack open my heart to let the light in. But the wilderness showed me that forgiveness is an oasis, where emotional thirsts can be quenched; and, as an oasis can bring flowers to a desert, so forgiveness can bring beauty to a desolate soul.

Sanele has never left me. He lives in me still, and I speak to him almost every day. I am so grateful to him for the time we had together, for he was a true guide. And I am so grateful to Baba Thabethe and the crocodiles – for showing me how to let him go.

———————•———————

I have had little material wealth in my life. But the wealth given to me by the wilderness can never be calculated. Nature has always been my medicine, my spiritual home and my teacher. As the crocodiles helped to show me how to free myself, so have many other creatures taught me much about how to live in this world. I have learnt time and again that nature brings the medicine we need to heal the brokenness in our world.

Yet each year, more people grow up having no contact with the wilderness. Technology and modern life dull our senses, and disconnect us from the natural world. Each day, new wilderness areas are bulldozed for farms, mines, houses and factories, more animals lose their homes, more species become extinct. We urgently need to understand how much we need these sacred places for our hearts, bodies and souls, before they are lost to us forever. My hope is that this book will inspire you to seek out nature wherever you may find it, to guard the last remaining refuges of wilderness fiercely, and help to create more spaces for wildness in this age of the Anthropocene.

Let us step into this story, then, and follow where nature will lead us.

Part Two

————•

A CHILD OF NATURE

1979 to 1998

VIVIDLY, STILL TODAY, I CAN SEE OUR FOUR SMALL MUD HUTS PERCHED on the side of a low, rocky ridge overlooking a seasonal stream. Known as uMjoyi for its salty water, my lonely little stream blanketed itself with wild ginger, wild lavender and berries. A tall igneous rock rose behind our homestead like a lighthouse. In the late summer afternoons, I would climb the rock and lie on the warm stone, watching my whole village during cooking time, as the sun went down beyond the undulating sea of hills. The smell of phuthu (maize meal porridge) floated over the village, mingling with the smoke from many small fires. The air filled with the bleating and bellowing of the goats and calves, galloping and playing with that singular unimaginable freedom of being alive and present only in the moment, with no shadows from the past or future.

On one such day, as I was watching the sunset, my heart suddenly filled with an emotion that felt too massive for my small body. There was such intense beauty in the world, the golden clouds scattered across the wide sky, the gleaming grass heads catching the last rays, the animals playing below; it was as if all at once I could sense the unbounded wonder of nature and life and earth. I wept because I did not know how else to respond, how to contain this feeling. I was still weeping when I came home – my mother begged me to tell her what was bothering me, but I had no words to explain it.

Our four mud huts stood precariously, undermined by the termites and woodlice nibbling the foundations, but they offered a sturdy enough home to my mother, my brother Siyabonga, my sister Makhosi and me – I was the last born. My father spent his weekends off with us, and my mother always opened our home to anyone in need, so there were usually a few extra sharing whatever food we had. My mother is a very strong woman from the Ndwandwe clan, a most powerful clan that dominated until King Shaka took over the Zulu monarchy.

In front of the homestead was a kraal for goats, and a vegetable

Goats at the Mbatha family homestead

My mother preparing umqombothi for a ritual

garden. Our fresh milk came from goats, as we did not have a cattle kraal until I was six. I remember being deeply moved by the first cow that my father bought. It seemed that the cows brought warmth and dignity to the homestead. We boys always aspired to owning cows, for we truly understood their value. Not only for milk – the oxen were used as a means of transport and for ploughing.

When darkness came, we'd gather after dinner around the dancing fire and embrace the elders' storytelling. These stories reinforced the lessons of ubuntu, the philosophical foundation of African community life, beautifully expressed in the saying 'umuntu ngumuntu ngabantu' – which may be translated as 'I am because we are'. Through ubuntu, communities build strong bonds that enable them to overcome hardships. Burdens and joys are shared; raising children is communal – we knew we could go into any homestead in the village to get food if we were hungry. Ubuntu helped our community to survive poverty through practices such as ukusikele (giving land for ploughing) and ukusisela (giving livestock). This acknowledgement of our dependency on, and responsibility towards, one another can only truly be understood in practice.

The stories carried messages of ubuntu: of the need to help one another, to work in harmony and to respect our elders. But they also stressed the need to respect nature and other life forms. We would drift in and out of sleep, the voices of the elders weaving through our dreams, carrying us deep into the essence of understanding, floating with us across the mighty rivers of our inner landscapes.

We had duties from a young age, but in those days of freedom before I started school there was plenty of time to mould clay cows by the stream, play traditional stick-sparring games with other boys my age, swim in the rivers, fish, and eat wild fruits. Sanele was my constant companion, and I followed him everywhere.

At the age of six I started my daily duties of caring for our herd of about sixty goats. Along with other boys my age, I learnt what I needed to know from an inqwele, an older boy who acts as a mentor to the younger ones. He taught us which plants and fruits were edible, what was poisonous and what was medicinal. We learnt how to keep our goats safe from predators, and how to hunt and skin wild animals. I remember when I caught my first guineafowl – I was with my brother. He took a white-spotted feather from the bird and threaded it into my hair, then smeared the blood on my chest. He advised me to kill animals only for my survival, never just for entertainment.

As a goatherd, I could spend all my time in nature. I was a small boy with innocent eyes wide open, and an untarnished mind eager to drink in everything around me. I learnt to interpret the calls of the bush, to cross both streams and mighty rivers (or so they seemed to me), to track the hours of the day by the sun's journey through the sky, to read the clouds and anticipate the weather. I came to embrace the valleys and mountains, to appreciate the sunshine for its warmth and the cold for waking up my senses, to rejoice in the exhilarating thunderstorms. I danced naked in the African rain under the shepherd tree, surrounded by the gentle bleating of my goats as they nibbled fresh leaves of the buffalo thorn. I discovered the varied textures of the earth's skin through my bare feet, from the rough, thorny bushveld to the soft, gold river sands. My days and nights were bracketed by sunrise and sunset, moonrise and moonset.

The gentleness in my heart was nourished by my daily responsibility towards my goats, as I learnt what it means to take care of others. I would swing my knobkerrie fiercely in the air, ready to take on any predators who wanted to steal my goats. I watched out for the martial eagles, soaring above and stalking the young goat kids; I looked out for the hungry jackals; I spoke to all the predators, asking them to find their meals elsewhere. When my goats were fearful, I calmed them. When cold winds blew, I gave them warmth by finding sheltered grazing land. I came to know each one, with their black enquiring eyes and dangling ears. I used my senses to commune with them and interpret their needs; I thanked them for providing me with fresh milk. Some days I became a kid myself, galloping about playfully and leaping from rock to rock, while the older goats became my guardians. The pure love that the mothers shared with their babies ignited my soul. I resolved to train myself so that when I was insizwa (a young man) I would understand the unknown languages of animals and nature. The goats, the plants and trees, the landscape around me, wove their way into my soul until they were all of me, and I was all of them. There was no barrier between me and the life beyond my skin.

My fierce protectiveness was not only given to my goats, but extended to all animals, and I hated to see an animal hurt. I remember once refusing to eat a fish I had caught, because the iridescent colours flashing on its skin were too beautiful – and I released it back into the water. One day, when I was a bit older, I came across a group of boys who had surrounded a baby rabbit, and were planning to kill and roast it. I could not bear to see this tender young thing suffer this fate, and fought fiercely for its life, nearly losing an eye in the process as we boys fought about it with sticks. The boys were angry to lose a meal, but eventually they gave in and let the rabbit go.

As the days passed, I felt stirring within me some vision of who I wanted to be. I imagined myself growing into a warrior, powerful enough to protect the spirit of the rivers, trees and mountains. I would climb the hills and the trees, and sit, feasting on the wild fruits, my mind spiralling with these thoughts as I gazed out over the treetops at the blue Zululand hills, rolling towards the horizon like an ocean without end.

I was to fight my first battle for nature at the age of six.

A few weeks after my sixth birthday, I found myself in a school classroom for the first time. After running wild like a young colt with only my goats and the hills for company, I found it shocking to be stuck in a room full of human kids for hours. We ended our first day with a dance while the teacher played a guitar, which lifted my spirits a little.

The daily journey of fourteen kilometres there and fourteen back was onerous for a six-year-old child. But Sanele encouraged me, and helped awaken my senses to the natural beauty along the way. In the early mornings, we would walk through the flowering trees, the air rich with the scent of the white flowers of the black monkey thorn or the sweet, red flowers of the weeping boer bean. Our music was the singing of the waterfalls, and the staccato calls of the bleating warblers in the thick, gold grass. We marvelled at the colours of the rocks, so bright under the clear, rushing water of the rivers.

Sanele helped me to find the best wild fruits to sustain us on the journey. The trees were truly our fathers and mothers along the way, giving us sustenance, and welcome shade on the blazing summer days, or shelter from the hard rains. Our hearts were uplifted by the majestic fig trees, and by the umbrella thorns that seemed to smile at us from distant hills, their arms thrown wide to welcome tired travellers and offer generous shade under their broad canopies.

But the queen of the trees was a great umbrella thorn that grew on top of a high hill near the school. The tree could be seen for kilometres around, and always seemed to encourage us as we made the long, hot trek up the hill towards it. When we reached it at last, we would flop down under its generous branches, grateful for the cool shade refreshing us for the rest of the journey. Often other weary travellers would be resting there, laying down their bundles of firewood or maize cobs to enjoy a few minutes in the shade.

One day, as we came up the hill, we could not see the familiar shape of the umbrella thorn. We hurried up the path, disbelieving our eyes, hoping it was some trick of the light and that the tree would be there as usual. But as we drew near the spot, we could smell fresh sap. We stood in stunned silence, gazing at the stump bleeding out its sap into the grey soil. The gold flowers lay strewn across the ground among the broken branches, while bees and butterflies fluttered about gathering the last scraps of pollen and nectar from the dying blooms. As I laid my hand on the wounded stump, a river of tears poured down my face.

A gust of wind sent the fallen leaves swirling around the stump as if to embrace the dying soul of the tree. I felt lost and bewildered, consumed by emptiness. How could something so meaningful be cut down so carelessly? How could something so generous be violated so cruelly? I felt as if I too had been struck down and violated. The bright memories of the stories we had shared under this tree flooded my mind. I'd only been attending school for a few months, but the tree had already brought me such comfort. In the way of small children who have witnessed little change, I had assumed it would be there forever, that one day my kids would sit under this tree, listening to me sharing the stories of my life.

But the tree had been murdered, and the place was enveloped in sorrow.

As we recovered from our first feelings of shock and disbelief, we children were galvanised by our anger, given courage to protest by the strength of our grief. Led by Sanele, we launched a peaceful 'protest' by writing messages saying things such as *The trees are us, who killed this tree?* We wrote these on rocks near gathering places, at river crossings, at the communal bathing places in the river for males and for females. Many of our parents were illiterate, but we knew that these messages would provoke their curiosity and they would ask us the meaning of them. Other kids joined us and the campaign grew, with messages appearing all over the area.

This action showed the village that we kids were committed to protecting nature from thoughtless destruction, that for us the trees were like parents, giving us fruits every lunchtime at school and sheltering us from the sun and rain. Our attachment to trees, rocks and grass kindled a spark of caring for the land itself, opened our ears to the voices of nature and urged us to harvest its spiritual benefits.

Our protests reawakened the dulled feelings for nature in our mothers, fathers and traditional leaders. It helped to revive their appreciation of natural beauty in our villages, to rekindle their delight in our trees and the streams with their small pools and tiny fish, the scent of the wild ginger bushes and wild basil. It reminded the villagers that, through all the hardships they endured, the natural features around us enhanced our lives and were true symbols of perfection. In response to our protest, traditional leaders laid down a law that all trees along the paths must be protected because of what they provided to kids walking to school and other travellers.

The tree had been cut down by a man who had spent some years in jail. At the time, I thought he was just an evil man, but looking back now I wonder whether those years in a hard place had so disconnected him from nature and his home that he had forgotten the trees' worth.

Through this victory I celebrated the empowerment of my connection with nature, the true freedom that comes with this connection. I drank in the life of the dying tree as if drinking from a stream with shimmering waters, and so I became its voice.

———————

School was a strange and foreign imposition on my life. It was so different from my life at home, my life out in the pastures with my goats. Walking so far without shoes was tough for a small child. The path was beset with dangers and, especially after Sanele was taken by the crocodile, it demanded superhuman courage to keep stepping into the muddy, churning rivers, never knowing what might be lurking beneath the surface. Apart from the dangers of crocodiles and fast-rushing waters swollen by summer storms, there was the ever-present danger of the water snakes such as amavezimamba. Other venomous snakes lay hidden beside the path, puff adders and cobras, and none more deadly than the feared black mamba.

I remember when one of the older men was bitten by a mamba in the village. An old woman sent us out to get the bark of the sickle bush. She ground it to powder and mixed it into a paste with other herbs that she kept in the household. She smeared the paste on the skin where he was bitten, and told him to keep still. The medicine absorbed the venom and stopped it from spreading too rapidly through his system – the patient remained calm and he survived.

This experience showed me that as much as nature kills, it also heals, for it was in nature that we found the medicine to help this man recover from the snake bite. But it was also a lesson in the power of our traditional knowledge. At school we were learning the propaganda of the colonisers and missionaries, which discredited all native wisdom. Yet it was indigenous wisdom, working with nature, that had saved this man.

Once at school, I felt lost and sad, as if my soul were being crushed by the confusing lessons, by the loud voices of angry teachers who would beat us with thick wooden sticks if we failed to understand. Sometimes children would have their fingers broken, or would have to go to hospital for their injuries from this punishment. The teachers would stand behind us while we were writing a test, and smack us each time we got an answer wrong. The classroom was crowded, with too many children to fit on the wooden benches. I was usually seated on the floor, partly because the long walk exhausted me so I often fell asleep and might have fallen off a bench had I sat on one.

In the weeks after losing Sanele, the daily journey was a grim ordeal. But in time I learnt to calm my fear and grief by listening to myself breathing, or by listening to the birds as I wound my way along the narrow traditional paths that scarred the virgin forests around my village.

At school I learnt to escape the torment of the classroom by going on imaginary journeys. I pictured myself playing in my stream, climbing the rock behind my homestead or moulding clay cows. I lived for the lunch breaks, even though I had no lunch box, for this provided an escape from the crowded classroom and a chance to embrace the coolness of fresh air. I would run out, delighting in the feel of the sun and wind on my face, in the wide blue sky stretching into the distance. I'd join the other children in foraging for wild fruits like a troop of baboons in the surrounding trees. When the school bell rang, I cursed it for taking my freedom, but ran into the classroom comforted by a stomach full of bush medlars, red ivorywood fruits, sour plums and water berries.

Going home was the best part of the day, especially in summer when these fruits were plentiful. We boys quickly learnt which trees produced the sweetest sour plums and other fruits – they all looked the same, but they tasted very different. We were 'connoisseurs' of the wild fruits, as discerning as any wine taster. The older boys, the izinqwele, shared information about medicinal trees and plants to treat headaches, toothache, cramps and diarrhoea, powerful organic medicine that nourished our systems. Our older sisters played a major role in advising us, and sometimes they would carry us to school on winter mornings when it was too cold for us to walk barefoot, or if we were weakened by illness. They would give us their pencils or pens when we lost ours. The older girls were the queens of this journey; the love, care and sympathy they shared were a great comfort after the harshness of the teachers. I learnt little that was useful in the classroom in my first few years, but I learnt a great deal on the walk home!

Deep down I knew there was something wrong with the education we were receiving. We seldom had the opportunity to learn outside, but I knew that all the greatest lessons in my life happened when I was outside. This was where I found my roots and strength, the courage to face the challenges of the day. In time I managed to develop the skill of learning and doing classwork quickly, so that I would have a few minutes to sneak outside, pretending that I needed the bathroom. I would go out and embrace the fresh wind, the rustling of the leaves in the trees, the high green summer grass rippling in the breeze. Sometimes, the air would be rich with the scent of rain. These breaks brought me such intense happiness that I was better able to tolerate the hours in the classroom, and so I found a way to pursue my education.

Sometimes, I was so desperate to escape the classroom that I would will the clouds to come together and create a lightning storm so that the teachers would let us out early. I didn't even think about the danger of travelling home while there was lightning; my only concern was to get out.

In winter I watched through the broken windows of my classroom as the trees cried leaf tears; I watched the shadows of the clouds moving across the hills, the patches of golden light gleaming on the grass when the sun broke through. Despite being crammed into a room with eighty other children, these sights enabled me to go deep into my own thoughts and escape the realities of the classroom. I could embrace this gentle life within myself by watching the dancing grass. This ability was more valuable than anything I learnt from the textbooks, for it enabled me to use those broken windows as a gateway to a lifelong voyage of self-discovery, to take refuge from the hard walls of the school.

Even though school was painful to me, as I grew stronger the journey to school came to fill me with profound tranquillity. The wild nature I encountered along the path, the different moods of the weather and the seasons, the small daily changes in the landscape all stimulated my senses, reminding me again and again to pay attention, to bring myself into oneness with my surroundings. I learnt to find courage, to trust myself to be brave enough and strong enough to face the hardships of life. It gave me time to reflect on and bond with whatever journeys my future might bring. It taught me about friendship – not only with my schoolmates, but about being a true friend to the trees, streams and rivers, and it strengthened my armour for the struggle to protect nature against those who were harming it.

These journeys deepened my resolve to be the voice of the trees, the rocks, the streams with sparkling waters and the rivers with muddy waters. I devoted my feet to bearing me into a world full of wonders, and devoted my soul to finding ways of awakening the hearts of my fellow villagers to these wonders. And so the journey to school became, for me, a source of love.

───────────

Among the most precious memories of my childhood were the times I spent visiting my father, who was working in the nearby Hluhluwe–iMfolozi Park named after the great White and Black iMfolozi rivers. 'iMfolozi' means 'the river of the stinging nettle'. The elders told us that, before Cyclone Domoina had hit the area in 1984, both banks of the river had been full of the plant called uluzi (the mountain stinging nettle) so it looked as if the plant was protecting the river. The plant was indeed a fierce guardian of the waters – it is very itchy when you walk through it, and it creates blisters and sores. But it was also useful and was harvested by the women, who used it to produce ropes. 'iMfula' is an isiZulu word for 'river', so 'imfula' and 'uluzi' became 'imfulawoluzi', shortened to 'iMfolozi'.

The Hluhluwe–iMfolozi Park is the oldest proclaimed nature

reserve in Africa. Its ninety-six thousand hectares encompass grass-covered hills, plains and rivers. Most people in our district have never had the opportunity to enter its gates. But I was lucky to be able to visit my dad in the park – and, later, to work there myself. Being a child at iMfolozi offered me a deep calmness and connection, healing me from the stresses of my daily life, especially after writing my year-end exams.

My father worked at the reserve for most of his working life, looking after the horses that were used for patrols. When I was there I would trot along after him as he went about his duties. A bakkie would pick my father up after his time off in town, and I would travel with him back to his compound in the reserve. I remember arriving once to find a horse called Nkanyezi (Star) waiting at his rondavel with her new foal. She had somehow known that he was coming, or had heard his voice as we were travelling there. He stroked the foal, praising Khanyisa for the foal's beauty and promising to look after them both. He encouraged me to stroke her too, saying, 'Don't be afraid, stroke her gently.'

Soon all the horses gathered around, nuzzling and whinnying for my dad. I thought it wonderful that a horse should come to show him her baby. I believe that my dad was a true and authentic horse whisperer – he talked to them, and they seemed to understand everything he said. Helping my father with his work helped me to develop a deep understanding and respect for these animals. I came to love them for their powerful energy, their wildness mixed with gentleness.

I remember the thrill of waking up to the sound of roaring lions not far from the compound; my father praising the lions, declaring them to be iNkosi Yehlathi – the king of the bush; the soft drumming of rain on the thatched roof of the rondavel; the ancient sounds of the night-time wilderness, the call of the hyenas and nightjars; how the horses in the stable would whinny and stamp nervously at the sound of predators.

I remember hearing the haunting cry of the African fish eagle for the first time; watching the slow circling of vultures high in the summer sky; the golden haze of the lush red grass. I recall watching a flock of great white egrets flying above the river, their pure-white forms perfectly mirrored in the still water below, seeing the reflection of my own face mingling with the reflected birds in the water. I remember the rich parade of life that was inscribed on the golden sands of the iMfolozi River – the journeys of spiders, tortoises, crocodiles, antelopes, elephants, the big cats and wild dogs, all recorded by their tracks, sometimes so fresh that you could still catch a scent of the animal. My father knew them all, and helped me to identify them.

After Sanele died, I was reluctant to visit my father for months as I was so terrified of the crocodiles. In time, I overcame this fear, and visited him again. But it would be some years before I could venture into the iMfolozi rivers, particularly the Black iMfolozi. The mud makes the water turbid, so it is impossible to see what is under the surface or how deep it is. I was able to cross rivers that were clear, but the opaque waters of the Black iMfolozi terrified me.

My father tried to calm my fears. But I remained frightened – until the day he faced down a pride of lions with his knobkerrie.

I was walking with him, looking for the horses. Down by the river we heard the tinkling of the bells my father had hung around their necks, but as we walked towards them we saw a lion pride stalking them. The horses had not been aware of them, but just as we saw them they caught the lions' scent and started stampeding straight towards us – followed by the lions. My father yelled to head them off, but in the commotion of ringing bells, pounding hooves and snarling lions his voice could not be heard.

Just behind us was a big umkhiwane tree that had fallen. When my father realised that the horses could not hear him, he lifted his knobkerrie, and began beating it against the stump of the tree. The stump was hollow, and emitted a loud bang when struck by the knobkerrie, like the shots of a handgun – pow! pow! pow! The lions were now only ten metres away from where we were standing, but the noise made them falter and run away. My father must have been terrified, knowing he had a seven-year-old child with him who might easily have taken fright and run – had I done that, the lions would have run me down and killed me. But he had the presence of mind and stoutness of heart to use what he had to stop the lions. This showed me that if you believe in yourself, you can face great dangers – that you have more power than you can imagine. After that, my fears receded and I felt safe walking with my father again.

My father always gave me space to lead when it was safe, and left me free to wander and explore. Having that freedom uplifted my soul and cleansed my mind of stress and frustration. I could wallow in the abundance of life around me, as a buffalo wallows in mud to soothe his scratches and tick bites. I developed a deep sense of connection and kinship with other species, as I foraged for wild fruits from same trees as the baboons and drank from the same pools as the lions.

It also galvanised my passion for a deeper connection with my own heritage, for the iMfolozi is rich in history, and in the history of the oldest tribe of all, the San people. Long before my Nguni forebears arrived, the San people lived here, and the place is filled with their rock paintings and stone tools. This was the cradle of all our ancestors, the land little changed from the days when they walked on the earth. I walked barefoot as they had walked, on the old paths that they had created, smelling the same flowers that they had smelled. This created in me a deep sense of belonging, and a profound gratitude for my heritage. I could drink from the well-spring of my origins, and feel the generosity and kindness of my forebears, the wisdom of those who had lived and died steeped in the practice of ubuntu.

I remember, one day, walking with my dad through tall red

grass. A cold wind from the south, thick with the scent of mud, was swirling around my knees. My dad murmured, 'Heavy rain is coming.' From far off we heard the tinkling of the bells around the horses' necks.

We sat under a fig tree, and I lay on my back to watch the clouds moving through the sky. Hypnotised by the whisper of the wind in the trees and the water splashing against the riverbanks, I drifted into a dream while my father softly sang a rain song. A Burchell's coucal sang along with my dad from the tall grass; the sound of the bells on the horses rose and faded with the shifting wind. I felt myself sinking into my surroundings, filling with the sensation that here was the true essence of life and I was immersed in it. I had no words for what I was experiencing, but it made me feel both shaken and grounded.

At that moment I knew, with a deep, whole-body knowing, that the essence of life is to walk in nature, to follow the clouds moving through the sky, to stay connected to the earth by touching the soil with bare feet and hands. This knowledge rushed into me with such power that I knew not how to contain it. I *had* to share it; I had to find a way of bringing others into connection with this wilderness, that they too could feel its power. Alone, I was too small to hold this power; we had to hold it all together.

'I want to be a game ranger!' I declared to my father.

I had but a hazy idea of what this was but as I uttered these words, it was as if the seed of my future life was planted in my heart. Young as I was, I knew that my life's purpose was to protect that seed until it could sprout. That if I took care of it, this seed could become a tree, it could grow into a forest, and it could grow into wildness. Little did I know then how hard I would sometimes have to work to protect this seed from my inner droughts when my courage failed, or from the toxic weeds of self-doubt. But I did know that I had no choice but to nurture it, for it was as essential to my soul as the air to my breath.

Early one misty morning in my sixteenth year, I woke to the trickle of rain running down our windows and walls. I could sense the cool stillness of mist enshrouding our small village, but my little mud hut defended me against the damp and embraced me with warmth. I lay listening to the baby goats complaining in the kraal, and their mothers comforting them. The smell of fresh smoke told me that my mother was already up and getting ready for her daily duties. A dog barked, and a young man rapped on my door, urging me to get up quickly as I was expected at my grandfather's homestead.

My biological grandparents had all passed away before I was born. But this man had stepped into the void they had left, and appointed himself as my umkhulu (grandfather). He was respected throughout the village, especially for his profound humility, and I truly loved him. He was strikingly handsome, with a black moustache and prominent eyebrows that seemed to add authority to his words. Sometimes, when he talked, he would drum the earth with his huge hands, creating a vibration that resonated with the rhythm of his speech so that it became a kind of song.

When I came into the hut, the dancing fire illuminated more than a dozen faces. There were about fifteen other boys my age already sitting there. I wondered what this gathering could be about – my mind flooded with fear as tried to recall any wrongdoing from the previous day, but I knew well that we'd come home from school and fulfilled our daily duties successfully. The intense silence in the room accentuated my fear, and my heart beat even harder when I glanced around the hut and noticed the bundles of sticks hidden under the goat hides. uMkhulu's moustache danced about as he greeted us with the words, 'Sanibonani, zinsizwa.'

We all looked one another and chorused, 'Yebo.'

uMkhulu continued, explaining why he had decided to invite us on this day.

So began my ukuthomba, my initiation day, the ceremony of transition from a boy (umfana) to a young man (insizwa).

The old man put his gnarled hand under the black furry goat hide, and pulled out a spear from the bundle. He rested its blade in the fire until it glowed red from the heat. He raised it into the air and spoke to us about being insizwa. Here I was, listening to a teaching of how to be sun, water, moon; how to be lightning.

'Be as the sun,' my grandfather said, 'for the sun brings life and warmth, and everything grows under it. It empowers even the smallest of the flowers. And be as the moon, for the moon brings light in the darkness, just as the scholars of the east bring the light of learning. And be as the rain, for the rain brings life and nourishment, and enables all living beings to flourish. And be as lightning, for lightning brings the strong energy we need for change; lightning brings the healing rain, and the fire that warms us and burns the old grass so that the new grass may grow; lightning can illuminate even the darkest moments of our lives. When you shelter from a storm in a cave, it is the lightning that will illuminate the cave, and show you in that single flash that there is nothing to fear in the darkness.'

He told us that the spear in Africa is a powerful tool that may be used to protect, to hunt for food, or to fight to save lives. He said that our bodies were transforming, but that physical transformation is meaningless unless our minds also got stronger, so that we could become good warriors who walk with peace and harmony on our path. His moustache danced like buffalo horns as he talked to us.

His words flowed like a fountain on our dry, lamenting souls, bringing coolness and hope to me and to my fellow youths. With the glow of the spear, he flagged the light amidst our darkness.

'Life has darkness, but even the darkness brings stars. Life has bright moonlight, but is also shadowed with clouds. Life has sun, but also cold weather. Life has both happiness and sadness. As izinsizwa you have choices, and you need to grow stronger in your minds so that you may choose wisely. Peace and harmony are the only spears to fight disease, poverty, inequality and hatred. Peace is the remedy for anger.'

He took another spear, and put it into the fire until it too glowed red. He told us that fire and water bring life, for without the sun there will be no life, and without rain there will be no life.

'As with the sun and the rain, our mind and bodies work together as a powerful force that can destroy the world, but that powerful force can be the medicine that heals the world. Together, our minds and bodies have the power to protect the village, the elders, disabled people, our country and the world. To be insizwa is to allow your soul and heart to be imbued with the wisdom of life, and give birth to the true warriors of harmony and peace.'

I saw the light from the sun creeping through the cracks of the old wooden door, illuminating the umsamo (the sacred part of the hut). I watched as the sunlight sneaked onto the faces of my fellow izinsizwa. The voice of my grandfather, the vibration of his fists hitting the ground as he talked, took my heart on a deep journey of mind and soul. My buttocks were complaining from having sat in the same position for so long, and my knees were numb, but my soul was flying like an eagle. uBunsizwa, the state of being a young man, was for me a journey to discover my new foundation and my own significance with humility and respect.

I was sixteen years old, not sure if I had reached the physical maturity of insizwa, although I had noticed some changes in my body. The words from my grandfather were important for people at all ages, but sixteen was a good age to acquire this status, founded on these rituals. I knew that among my peers such initiation rituals were being replaced by baseball caps, alcohol and parties, which they believed could show their status as young men. The initiations were stigmatised as being associated with witches and godlessness, and were being lost to our people.

I knew that I was discovering my own roots through the words of my grandfather, that his words were healing rain for my thirsty soul, a lightning flash of truth in a moonless night. But though his voice was strong, I could hear sadness, the cry of an old man watching the dying of his way of being, for he knew that without these initiation ceremonies we would be lost souls, easily blown away like trees without roots.

His gift of a spear and tail cords from the bull made us realise that we must also do ukugwenda – a form of traditional circumcision that causes the foreskin to retract. This ritual was already dying out, and I was one of the last candidates who embraced it in my village. It is sad to know that my own kids may miss this important part of life's lessons. Nowadays, young men just get promoted to being a cow herder with no such meaningful talks. Life in rural areas has been uprooted by modernisation, and we were blessed to have heard my grandfather's words and taken part in this ceremony.

———•———

Early on the morning after my ukugwenda I walked into the cattle kraal to begin my new life as a cow herd. The dust of the kraal gleamed in the dawn light. The air was filled with the bellowing of the calves and cows, and my brother's piercing whistles. As we went down to the kraal, the older boys sang our praise songs to welcome us as izinsizwa. My heart lifted to hear my own praise song: Shlahla somqumo esigungwe yizizfiki, cembe lomviyo eludliwa zimbuzi zabanumzane, inkonjani ekhala igijima iyongena kwaNxumalo. The song talks of a wild olive tree that is embraced by visitors – a reference to a time when I was invited to attend the birthday party of a daughter of a white section ranger; it sings of leaves of the bush medlars that feed the goats of a respected man – referring to the love between my grandfather and me; it speaks of

Siyabonga milking a cow in the Mbatha homestead kraal

the swallow that chatters in the Nxumalo homestead, for I spent much time in the home of my mother's brothers.

The cows seemed to bellow in response, and together the voices rose in a chorus of harmony between human and beast. As I sat down to milk, guided by my older brother, the warm breath from the cows dispersed into the chilly air, and I was grateful for the warm teats against my cold hands. Flies were buzzing around my ears, disturbed by the flicking of the cow's tail as she chased them from her flanks. The foaming milk thrummed into the container between my knees with a soothing rumble as it slowly filled up.

The kraal is a sacred place in my culture. Cows are not just animals but sacred beasts, symbolising and enabling the connection between the physical and spiritual worlds, for they are the means by which to perform spiritual rites. This understanding was planted in my mind by uMkhulu during the talk at my initiation. I understood from him that herding the cattle is therefore a sacred duty, for you are tending not only the family's wealth and prosperity but also the living beings that connect us to our ancestors. As I led the animals out for the first time, I felt a shiver down my spine at this responsibility.

Proudly I walked behind the herd of amaqanda ka huwe – cream-coloured cows spotted lightly with rust. I stood in the warmth of the soft African sunshine, watching the smooth flight of a black-bellied bustard as it came to land among the peacefully grazing herd. The bellowing of the cattle was answered by the echoes from the hills. A Cape turtle dove was singing from nearby trees, its song mingling with the calls of the other birds. As the day drew on, the cow udders became so distended that the milk started to drip on the ground – what we call ziphakela amadlozi, or feeding the ancestors. The comical antics of the young calves made me laugh as they gambolled across the grassy pasture; the young bulls sparred with one another, just as we young human boys did with sticks. I felt myself growing, within and without.

During my days out in the hills with the cattle, I was struck by the simple happiness of the animals. The young calves showed such pleasure in running about, sparring, drinking from their mother. It made me wonder why my own life sometimes felt so incomplete or difficult. We heard our elders talking of their problems, and often felt burdened by the many hardships of the world. But the cattle helped to show me that happiness can be found in the simplest things, if you allow yourself just to be present in that moment. Happiness does not depend on great wealth or luxury. It is always available, depending on how you seek it. Watching the joyful freedom of a gambolling calf, or horse, or dog – that absolute delight in their movements and their bodies – brought joy to my soul too.

Of course, there were days when it was cold, when storms came, when the heat was intense, when the wind blew. It was my responsibility to stay with my cattle, no matter what was happening. But I thought of my grandfather's words – embrace the rain, for it symbolises abundance, the wind, for it symbolises transformation, and the sun, for it symbolises life. This became my mantra, enabling me to see even difficult times as an opportunity to grow.

Like the young bulls, we izinsizwa also did our sparring, using sticks in traditional stick-fighting or ukuncweka. We would break flexible sticks from the trees and spar with these – this kind of sparring was more like training or play-fighting. But more serious fights sometimes broke out between boys from different villages after school. These matches were physically very tough, and sometimes caused injuries. They became so serious that adults got involved in the disputes, and one of the fathers was shot. After that, the elders banned stick fights at school, although we still used to bring our sticks to school and hide them so that we could defend ourselves if boys from other villages attacked us. Stick-fighting taught me to stand my ground and fight for what I believe is truly mine. It helped me to learn to defend myself, and honed the skills I would need to defend my family in the future.

As insizwa, I could also now participate in important rituals and ceremonies such as burial ceremonies and amahlambo (cleansing ceremonies). Now, I was invited to listen when my father praised our ancestral spirits during rituals for these spirits at home. Learning and understanding the names of these ancestral shades showed me how to commune with the divine ones, and how to interpret their language and their signs. This learning is essential for strengthening the relationship between living family and its ancestral guardians. It would enable me to play my part in protecting my loved ones from misfortune and safeguarding the family's eternal journey, the unbroken connection linking our ancient antecedents to those who were yet to be born.

———

As I grew towards manhood, I strengthened my resolve to be a warrior for nature, a game ranger who could bring the wilderness to others. I tried to express this to my friends, but they couldn't understand it. They couldn't believe that, after spending so much time in the bush as a cow herder, I was dreaming of spending yet more time in the bush. I could not blame them, as the apartheid regime's 'Bantu education' did not encourage us to work with nature. In fact, they would make us work in the school garden to punish us if we were late for class. This taught us that working with the soil was punishment, something undignified. Bantu education encouraged people to become teachers and nurses, clerks and police officers. I never heard any teacher talk about game ranging, even though we lived just a few kilometres from the game reserve.

My friends had different ambitions in life. They liked big cities and fancy clothes. For them, the signs of success were flashy Rolex watches, cars and luxury yachts, mansions with swimming pools, private jets, a young woman by your side laden with Christian Dior and Armani handbags. They delighted in looking at pictures of these things and discussing how they could acquire them, but I did not trust this hunger for material goods. It felt like a delusion, one that, in the end, could destroy the happiness and peace of mind of those who succumbed to it. To me, these symbols of status and wealth lacked the honesty and consistency of the natural world.

I was the same as these young boys – we were all birds of a feather – but I was flying in a different direction. I was the only one who had wheeled away from the flock to seek a different way of being in the world.

As my mother watched my growing determination, she encouraged me to follow my true path, to take whatever time I needed, and to be patient. 'The most important thing,' she always told me, 'is to be satisfied with your life. Life is like baking a loaf of bread. You must give it time to rise and to cook properly. If you cook it before it is risen, it will be hard and flat; if you eat it before it is cooked, it will be doughy in your mouth. When you are in too much of a hurry to chase life, you will not be satisfied with what you find.'

She would point out the young men in the village who had left school early to find work on the mines – some had failed, and turned to crime; they came back to the village with money in their pockets but anger and shame in their hearts, spending their money on alcohol to help them forget their troubles while their children went to bed without food. The young adults in our village were impatient for wealth but, as I grew older, I could see for myself how this impatience seldom brought happiness to the homestead.

When I was sixteen, I went with my school class to visit a big city for the first time. As the bus drew away, clouds of dust swirled behind the massive tyres that crushed the innocent soil. I looked through the bus window as it roared past my family's buffalo thorn tree, setting the leaves and branches trembling in its wake. The goats grazing under the tree scattered in terror of the growling engine. An old man walking barefoot in threadbare clothes waved us goodbye. When we crossed the bridge, I saw the flash of white on a spur-winged goose as it dived under the water. I looked back at my homestead, which was slowly vanishing, as if it were being swallowed by the dust cloud behind the bus.

When we turned onto the N2 towards Durban, I sat by the window, amazed at the steady stream of trucks and cars going by. My mind was filled with images of what this city might be like, and how different it might be from the village. We had heard so much about this wonderful city life. At school, teachers were always tantalising us with the opportunities in the cities: how city life offered us riches, and a better, easier way of living. This was the life that most of my peers were craving. I already knew that the path to my future did not lead to the city, but I was insatiably curious and excited to see it for myself. I wanted to embrace the experience for everything it could teach me, even if it showed me a different purpose in life.

The teacher acted as a guide, bellowing information with the insistence of a commander instructing his troops during a war. It was easier to pay attention now, thanks to the smooth ride on the tar, so different from the bumpy jolting of travelling on the roads around my homestead. I felt enveloped by sadness when our teacher told us that the highway we were travelling on had once been a major route for elephant migrations from the Cape to Mozambique, before these gentle creatures were slaughtered by the colonists for ivory. Although our village was so close to the game reserve, I was the only child on the bus who had seen a live elephant – some had not even seen photographs, and were confusing them with dinosaurs or rhinos. As the teacher tried to explain to them what an elephant was, I closed my eyes and tried to picture this route when the elephants used it. I imagined their waving trunks and flapping ears, heard their squeals and the soft tramping of hundreds of elephant feet. Their ghostly forms seemed to drift along the road beside us.

When I opened my eyes, I saw no elephants, only a parade of pantechnicons carrying coal, timber and sugar cane, and endless sugar cane hillsides, interspersed with sprawling eucalyptus plantations. Here and there, a few isolated pockets of indigenous coastal forest remained standing. They looked so lonely, dwindling islands of wilderness in an infinite ocean of cane. At the time, I did not fully understand the ecological damage caused by monoculture. But I knew from my time in the iMfolozi that these crops did not belong on these hillsides, that they had driven out thousands of plants, birds, insects, snakes, predators, antelopes …

My heart cracked at the loss of so many living things. Surely there was a way to live off the land without destroying life like this?

When we crossed the Thukela River, the teacher told us about the ultimatum tree. It was under this tree that representatives of the Zulu king, Cetshwayo, received an ultimatum from Sir Bartle Frere that led to the Anglo-Zulu War in 1879, and the ultimate crushing of the Zulu by colonial forces. Cetshwayo was reluctant to embark on a war, but Frere was determined to provoke conflict. Without the authority of Queen Victoria, Frere presented Cetshwayo with a set of conditions that he had devised, to which no sovereign king could agree. These included disbanding the Zulu army and swearing allegiance to the British queen. On 11 January, one month after presenting the ultimatum, Lord Chelmsford led about eight thousand British soldiers across the Thukela in an invasion of Zululand. The British suffered heavy losses in the first battle at Isandlwana, but their guns and cannons ultimately defeated the spears of the Zulu impi in the last battle at Ulundi in July. Cetshwayo was forced to flee, and was later arrested and imprisoned in Cape Town.

I listened to his words, while gazing at the broad Thukela River that once divided the colony of Natal from the Zulu kingdom. Between the lush reeds the water gleamed silver, making me think of the sun glinting on the raised spears of the impi as they ran in formation to Isandlwana. The original ultimatum tree, a giant wild fig, had been damaged by floods and fire. Now, only a decaying branch was left, holding out its dead arm and peeling bark like the Zulu soldiers surrendering to their defeat. So was the land of my ancestors lost to a foreign nation. As the bus crossed the river, I gazed at the reflection of the dead tree in the river, thinking about how the hunger for power has ruined the world.

When we entered the city of Durban, the bus was enveloped by a cloud of grey haze, bringing the smells of tar and petrol, and the greasy scent of the pizzas and burgers cooking in the fast food outlets along the way. The pavements were forested with brightly coloured traffic signs rooted in the concrete; people of all shapes and sizes – young and old, smartly dressed and in rags – were rushing in different directions. The air was filled with a cacophony of hooting cars. A salty sea wind grew stronger as we approached the main city.

Our bus pushed its way through the traffic, switching from one lane to another, trying to compete with the cars. I stood up at the window, trying to take in all these strange sights and sounds, amazed at the dense tide of humanity that flowed along the streets, the towering buildings blocking out the sunlight. The place was pulsating with energy, yet some parts seemed soulless to me. As the bus drew to a halt, our teacher bellowed with a voice like a dying horse: 'Okay, children, welcome to Durban, the biggest city in KwaZulu-Natal.'

We tumbled out of the bus and onto the pavement. I walked along behind the teacher, feeling the hot tar under my bare feet vibrate from the running engine of a huge garbage truck parked on the other side of the road. A filthy scent hung in the air. We walked past so many desperate people – a young man begging for a piece of bread, an old man at the traffic intersection begging for change, a woman with kids begging for food. For the first time in my life, I had to walk past someone in need without offering help, and ignore their heart-rending pleas. We had no beggars in my village – if someone was hungry, others would offer them food. Here we were walking past shops and restaurants laden with food, and yet the streets were filled with people who seemed to be starving.

I stood on the burning pavement, feeling sad and overwhelmed. The city was not the palace of riches our teachers had promised. It seemed to me to be a place of sorrow, a symptom of a world turned upside down, where food was thrown away on one side of the street while people starved on the other. From the outside, it seemed to have the flashy colours of a can of Coca-Cola, but when you drank it, it was just sugar and emptiness. I thought of the dead sugar cane fields, the ragged people begging for food. What had gone wrong in the world, that humans were destroying our earth mother to grow crops that people could not eat, while people walked hungry on the streets?

<hr>

After the trip, many of my schoolmates spoke of their desire to go to Durban to find work and lead a city life, but I became even more resolved to remain true to my calling. A few weeks later, my class teacher asked us about our dream career. I announced firmly that I wanted to be a game ranger and take care of wild animals. The teacher looked surprised, but she nodded and said, 'I think you will one day achieve your life's vision.'

I knew that game ranging work paid little, that it would never be a passport to the big car and the big house. But the seed that had been planted when I looked after my goats had been nurtured by my walks to school, by our fight to defend the umbrella thorn, by my walks with my father in the iMfolozi, by the words of my grandfather, by my visit to the city ... every experience in my life had nurtured that seed and pushed me more firmly onto the path to the wilderness.

I was not bothered by my friends' incomprehension. If anything, not being able to explain my calling to my friends showed me that you cannot learn about the gifts of the wilderness from a book, or a lecture, or by someone telling you – it is something you have to experience for yourself. I wanted to create a gateway for others to experience this. I wanted to protect 'my' game reserve, not only so that it could continue to nourish wild animals, but also so that it could nourish human souls.

I passed my matric with exemption, which meant that I had the qualification to go to university – but I did not have the means. Realising that my parents could not afford to send me to university for nature conservation studies was a bitter blow. Black youth from rural areas know all too well how frustrating it is to be a child with a desire to shape your future and the knowledge that your parents cannot support or assist you in doing so.

As young men, we were all under pressure to earn money so that we could help our families, and pay lobola. This gift of cattle or money is traditionally given by a man to his fiancée's family to thank them for raising and caring for his soul partner, and to pledge his commitment to support his wife and family. It is not a 'bride price', as people cannot be bought; this understanding corrupts the traditional meaning.

We needed to raise lobola to be able to get married. Yet as we grew into manhood we realised how little opportunity we had, and this led many to despair. There were few positive stories to inspire us. We all knew of young men from my village who'd travelled to Johannesburg to search for work in the factories or the mines, who'd drifted into gangs when they could find nothing. Some of the brightest boys, with distinctions in maths and science, landed up as hitmen (izinkabi) for the taxi industry because they had no opportunities to study further and build careers. They killed rivals for a taxi boss, and were rewarded with a fee and a taxi route. They came back home flush with cash, paid lobola for a wife … but now they were stuck on this road, forced to kill again to keep their new wealth and power, or because they would be threatened by the taxi mafia if they didn't. Some ended up in jail; others committed suicide when they knew the police were closing in on them. The women they'd paid lobola for were left as widows, with children who had no father.

Life in rural areas is like climbing a mountain barefoot and without food or a map. Only a handful reach the summit: many get lost in the valleys, others fall and never find the courage to climb the mountain again. All my friends had ambitions for well-paid jobs in the city. Only one succeeded.

I was determined to resist these dead-end choices, and do whatever I needed to climb the mountain and follow my path. I decided to work as a volunteer for Ezemvelo KZN Wildlife – the governmental organisation that managed the Hluhluwe–iMfolozi Park and all other national parks in KwaZulu-Natal. I knew that working for free, without even a stipend, would demand much courage and fortitude. But it felt like I had no option. I could never work inside four walls. I had to find a way to work in the wilderness, even if my family could not support me with food and I had to go to bed hungry. It was better to do something that might help me on my life's journey than just to sit, waiting, wishing.

So began my life's next chapter.

Part Three

———•———

STEPPING INTO THE WILD

1999 to 2002

THE SUN BEAT DOWN ON OUR BACKS. I HAD NO WATER BOTTLE, FOR I didn't yet know how important this was. The air was still, the cicadas shrill in our ears. The heat slowed us down, and with dry throats we retreated to the river and drank from it. My patrol leader undressed and waded into the water, and I followed suit. All this was done in silence.

After a swim I strolled around naked, studying the animal tracks in the sand by the river. Walking naked enabled me to dive into a sacred place. With no layers between me and the world beyond, I could feel at one with my surroundings, feel both the simplicity and the depth of my body's connection to the earth.

This was my first patrol as a volunteer. My path into volunteering had been smoothed by Sabelo Msweli, a section ranger who had known me since I was a child from when I used to visit my father. He and a white ranger called Craig Reed, who also knew me, encouraged management to take me on. Sabelo was like an uncle to me, and provided much support during my time as a volunteer.

When I started, I had renewed hope of being able to study further. Because of my good matric results, my father had undertaken

With Sabelo Msweli

to retire in a year or two and use his pension money to pay for my university studies. I had already been accepted to study nature conservation at Mangosuthu University of Technology, but I decided to volunteer until my father could pay for my studies.

This turned out to be further away than I could ever have imagined.

My home for my first six months as a volunteer was in Mduba compound, located in the Mduba hills. Our compound was surrounded by huge boulders, which offered an expansive view of the undulating hills of the iMfolozi. It was far from any signs of 'civilisation', with no light or noise pollution, and the view of the hills gave the true feeling of being in the bush for they stretched out as far as the eye could see. It was surrounded by groves of scented-pod acacia – the pods are a delicacy for antelopes, baboons and elephants, and these animals were frequent visitors to our camp. The unique sweet scent from the flowers attracts bees, butterflies and sunbirds.

We were accommodated in a modern building, consisting of a row of adjoining rooms with outside doors. My room was actually a spare room used to store equipment like lawnmowers, slashers and rakes – when I arrived I had to move these to another storeroom to create space for myself. There were no curtains, but I didn't bother to try to find any as my back window looked out onto a beautiful large-leaved rock fig tree, hugging the boulders with its gnarled roots. I kept my few clothes in an old suitcase I'd found at the school dumpsite. Here I had my first experience of a bed – an ancient iron contraption with rusty springs that squealed like an old pig whenever I rolled over. I had a small coal stove, which I seldom used as I usually ate with Dumisane Khumalo, another ranger. My first visitors to my new home were the scores of geckos that clung to the walls, feeding on the many insects. They became good friends – I named them and was pleased to see them at the end of the day.

I shared this compound with Dumisane and Baba Thabethe. Dumisane had been almost like a son to my father, although he was quite a bit older than me. When I became a volunteer, my father asked him to watch out for me, and he took me under his wing, giving me food, clothing, companionship and helpful advice. When he went on leave, he told me to feel free to use his house, and he stocked up with food for me before he left.

Baba Thabethe was a tall, powerful ranger from Sodwana, tough as nails but with a good heart and much wisdom.

Later in my time there I would participate in many of the different tasks in the game reserve: daily foot patrols, helping researcher teams to capture data, monitoring animals, notching black rhino, branding buffaloes, doing lion call-ups. But while at Mduba, I was mainly doing daily patrols and observation posts to monitor poachers.

The weeks and months ahead would be tough – working without pay, without a uniform and, on some days, going on long foot patrols without food. I was able to endure it only through the help of the rangers, in particular Dumisane and later Nkalakatha Nxumalo. My friends often asked me why I chose this suffering. But for me it never felt like a choice. From that first day, when I studied the patterns of animal tracks in the sand, stepping into the wilderness felt like stepping into a great way of being, into a tapestry woven from layers and layers of patterns. I knew that I needed to find a way to weave myself into this tapestry, to become one with all those layers. I had to follow those patterns, just as water follows the shape of the river, as a butterfly follows the wind. And so, tough as it was, working as a volunteer showed me again that my purpose in life was to work in nature, and it gave me the courage to follow my life path, however difficult.

One of my most profound encounters came through a mark left by an animal in the sand, a few weeks after I'd started volunteering. We were on daily foot patrol when we came across a fresh lion spoor, perfectly outlined in the soil, and fifteen centimetres across – what a massive paw must have made this mark! I knelt and placed my right hand over it. As my fingers touched the spoor, I felt a tingling vibration run through my hand, up my arm and into the centre of my chest, as if I'd touched an electric wire. I jumped up, alarmed. My companions asked what had happened, but I couldn't explain it. This was a message just for me, a personal telegram from the wilderness.

After feeling the power of the lion so vividly, I had to dig deep to find the courage to keep tracking it, but Dumisane and Baba Thabethe insisted that I should meet this beast. We kept following his spoor, clearly outlined in the sand, each pad like a teardrop on the earth's skin. All his actions were imprinted on the sand – we could tell where he slept, where he sat, where he drank, where he trotted and where he lay down.

Suddenly, he was before us. No longer just the memory of himself written in the sand, but living flesh and blood, fully and terrifyingly present, lying with the pride on the river sand. When he spotted us, he growled softly as he gazed at me with fiery eyes, burning into my heart. I felt, again, that tingling shooting through my arm. My patrol leader froze and told me to stand still. I watched, all my senses alert, as the lion twitched his tail from left to right. I was consumed by sensations of ice and fire. The lion looked me straight in the eye, his own eyes flames that kindled sparks inside me. I could feel his hot energy, crackling through the grass and into my body. I could feel the blood pumping through my heart, my lungs drawing in air … I had never felt so terrified, or so alive.

We watched until the lions moved away and disappeared into the thick grass. Then we walked back in silence, each quiet in our thoughts.

This was my most powerful experience with an animal spoor. But I had always been fascinated by animal tracks, these messages in the sand. I had learnt from my father that to a skilled tracker, tracks can reveal details of the animal's strength and health, whether it was running, trotting, meandering or limping. Everything leaves a trail. Even the tiny details of a feather on the ground can be revealed in a track, as can a grass head blowing against the sand. Tracking and intuition go together – trackers might lose a trail, but their intuition will show them where to pick it up. It is also an exercise in patience, for if you hurry you could miss an important detail.

Later, I would learn from one the best trackers, Nkalakatha Nxumalo. When he tracked an animal, he would take on the gait of the animal. When he lost the spoor, he would know exactly where to go – it was as if he was tracking with closed eyes but an open heart, his hidden senses wide open.

From him I learnt that following a trail creates a sacred space for you to become one with the creature you are tracking, and to become a true wild creature. It creates a 'here, now' moment, where nothing but the trail exists. You are pulled so deeply into the animal you are following that you lose all awareness of yourself, and move almost without realising what you are doing. By losing yourself, you discover the wild animal within you that you did not know existed – the animal self that flows into the earth, air and water around you.

Nkalakatha also knew intimately the scent of each animal. I remember him once trying to explain the scent of a leopard. I sniffed the wind, but could smell nothing at first. Then I caught it – a smell something like roasted peanuts. We followed the scent, and there she was, a female leopard in an old den with her cubs.

That night, after tracking the lion, I fell asleep on my squeaky bed with feelings of great accomplishment. I was still not being paid and had little to eat, but I knew I had experienced something unique and profound, beyond anything money could buy.

Touching the lion spoor

———

The strong smell of the elephant's urine assailed my nostrils as we walked, perhaps twenty metres behind him. Flies, butterflies and dung beetles were feasting on his fresh dung. Baboons were shouting from nearby trees as they gorged on marula fruits. We walked slowly behind the great beast – not tracking him, just happening to be sharing his path.

I was on daily patrol with Baba Thabethe and Dumisane. I had been volunteering for a few weeks, and was getting used to the routines of the daily patrols. But it was never repetitive, for each patrol brought new wonders, and new learnings.

The elephant carried on feeding, snapping trees and branches as he moved through the area, undisturbed by us. When he rumbled, the sound vibrated through my own stomach – I felt it as if I was part of him. Why could I feel this, I wondered? Would any human feel that sound? Or was it because I was so open to this connection? Was it that I did not consider myself superior, or dominant? That I felt as if I was one with this creature, sharing the same path, sharing the space and sharing the air we breathed? Would I have experienced this had I looked on that elephant with arrogance? I wondered how humans had come to imagine themselves as superior to any other creatures on earth. I knew that this way of thinking had set us apart, sent us drifting away from the sacred connection with other living beings.

I wondered whether it was the white settlers who had brought this way of thinking. From the stories of the elders, it seemed that the indigenous people had lived more peacefully with the wild animals before the settlers had come with their guns. The elders in my village had told me stories of how white rhinos would graze with the cattle around our homesteads even a few decades ago, but nobody was ever attacked or harmed by them. They told stories from their grandparents of hundreds of buffaloes herding peacefully within our lands.

Many of the Zulu clans are named for a totem animal – the totem of the Nyathi tribe is the buffalo; the totem of the Ndlovu tribe is the elephant; for the Dube, it is the zebra; for the Ngwena, the crocodile. These animals were acknowledged because they were so important to our people, providing food, clothing, ritual garments, medicine, and also helping to shape our identities. Traditionally, that tribe was responsible for the protection of that animal – if you wanted to kill a buffalo, you needed the permission of a member of the Nyathi tribe.

Some clans did not have totem animals, but were deeply connected to trees or features of the landscape. My own Mbatha clan is very connected to mountains, from a time in our history when a mountain helped us defeat King Shaka's warriors. The surname Manele means 'man of the valley' – a person of that clan would never allow the valleys to be destroyed by digging or mining. But as our traditional beliefs became diluted or suppressed by the missionaries and the colonists, so this sacred connection weakened or broke, along with the protective power of these clans over those animals or natural features.

When the white hunters came with firearms, they would easily kill a hundred buffaloes or antelopes in a day – far more than they could ever eat – and just leave them lying in the veld. The elders told us stories of animals found dead with bullet wounds, some with old, infected bullet wounds full of maggots – they had walked for weeks in agony until these wounds had eventually killed them. Finally, the white government began to round up the animals and fence them in, for it realised there would be no more to hunt if it did not preserve them.

The white settlers also treated the indigenous people, and the land, with cruelty and contempt. Indigenous people were forced into labour, driven off their land. Sacred rivers, lakes and forests were invaded and violated, destroyed to raise profits from farms, logging and mines. The settlers had no relationship to the land; it was just a resource to be used and abused.

As we walked the path of the elephant, my soul craved those times when people and animals had shared these ancestral lands in balance and harmony, without destroying one another. I longed to roam free in a land where the vast breeding herds of springbok could migrate freely across the plains of the Karoo; where the buffaloes could wander the grasslands with no fences to stop them; where the elephants could tramp from north to south without fear of being hindered or killed. I thought of all the fences, marching across the lands, chopping up the ranges of the animals, blocking their age-old migration routes, squeezing them and our people into smaller and smaller corners of the land they once roamed freely.

We reached the river and watched the elephant bull as he began digging for water in the dry riverbed, scratching up the earth with his foot. He drank for long time, surrounded by baboons, impalas, zebras, guineafowls, waiting around or chasing each other play-fully. Seeing these different animals at peace created a deep easing of my heart that calmed the agitation in my soul. It seemed to promise a different way of being, to show that the world was not only a place of strife and greed, of fences and oppression and control, but also a place of sharing and collaboration. I already knew that walking in nature could bring healing to my own emotional hardships, but I began to wonder whether the wisdom of nature could help to heal the sorrows of the world.

The elephant bull drifted slowly towards the grass and the animals came forward to share the water from the hole he had created – in this way, the elephant had brought water for all the animals. Baboons drank with impalas, without interference. Hadeda ibises flew above us and landed near the waterhole; cattle egrets followed the elephant and chased the insects he disturbed as he walked through the grass. So many animals were supported by this one elephant bull – the dung beetles, butterflies and flies feeding on his dung; the animals drinking from the waterhole; the ibises and egrets ... How unselfishly all the animals bring benefit to one another, how peacefully they share resources.

A cool wind took our scent straight to the elephant. He froze. He raised his trunk and faced us. I looked at him, and spoke to him from my heart: 'It is all fine, big brother ...'

He rumbled softly, as if in reply, then slowly and calmly began walking towards a thick forest. He had read our peaceful intentions, as we had read his.

Our two-way radio crackled with distorted voices – other rangers were telling us about a breeding herd of elephants coming in our direction. Soon we could hear them, trumpeting and rumbling, and the cracking and splintering of the trees as they fed. Clouds of dust swirled in the air above the bushes as the herd came into view, walking towards the dry riverbed. Two big females led the group. Young calves were dawdling, swinging their trunks from side to side, or playfully chasing guineafowls. Subadults were sparring in the riverbed, the clash of their tusks ringing through the stillness.

I watched, awed by their discipline, the complexity of their relationships. How could people kill these animals just for their tusks? How could anyone be willing to pay vast sums to decorate their living rooms with coffee tables made of an elephant's foot, or with trinkets carved from ivory? In what world can the value of such wise, extraordinary creatures be measured only by their body parts?

We walked on through air sweetened by the sharp, fresh scent of the peeling bark of the sweet thorn, and the white flowers of the black monkey thorn. We walked in silence, avoiding the females with calves. Behind us dark clouds were gathering, announcing the coming storm with thunderous rumbling across the savanna. Hamerkops and Burchell's coucals were singing for the rain. Heavy drops began to fall, accompanied by cracks of thunder and lightning. From afar on the ridge the zebras were calling. The drops turned into a downpour, and soon we were drenched.

I was pleased to reach our compound at last. The drops were drumming on the roof of my hut as I went in and opened the window, and whistled for the African rain. I took my journal and sat on the small veranda, recording my impressions of the day. Tiny rivulets of water were streaming on the ground; earthworms, giant land snails and millipedes were cruising around. The approaching night was heralded by the songs of crickets and frogs, and I was soon enveloped by darkness. I gave up writing, and sat quietly, inhaling the rich smell of the earth when it embraces the rain, listening to the rumbling of thunder fading away as flashes of lightning far on the horizon illuminated my surroundings.

One morning, we walked on a path created by wild animals that took us along the edge of the ridge, making our way to where we would set up an observation point to watch for poachers breaking the fence to hunt illegally in the game reserve. Soft clay soil, drenched from the recent rains, caked beneath my boots. As my boots got heavier from the mud, I laughed at my clumsy attempts to walk.

The sun rose higher, drying the mud and burning our faces. Sweat stung my eyes as I swatted at the flies around my face with my grubby old khaki cap. As we walked in silence towards the stream, the sound of the water gushing on the rocks made my heart leap with happiness.

I flopped down on the damp soil and opened my ears to the gurgle of water, the buzzing, whirring and chirping insects, the high twittering call of the paradise flycatchers. I was startled by barking and jumped up, alarmed. When I lifted my binoculars, I discovered that the sound was coming from a group of nyalas browsing on the far side of the stream. I was astonished that an antelope could make such a loud noise!

We set up our observation point near the stream. When you are doing an observation post, talking and smoking are discouraged, especially at night as the glow from the cigarette and the scent of the smoke can easily warn poachers. So the observation posts offer a perfect opportunity to sit in silence and contemplate your surroundings.

As I sat quietly observing, I felt my deep connection to the earth beneath me, as if the land were flowing through me. When I laid my hand on the soil, it felt like a warm, living thing, an extension of my body. This attachment to earth felt like a good, healing attachment, because it was to something life-giving and enduring, to something that brought benefits to all creatures who dwelled on it. It was different from the attachments to status symbols like big houses and cars. Most of my age mates longed for these, but I wondered if getting them would satisfy their hunger, for wealthy people so often seemed to want more. I thought about my trip to Durban, about the lost elephants, slain in their thousands for ivory. I thought about the mines built in sacred places because of the attachment to wealth ... how had the attachment to material objects and status become so strong that humans were willing to destroy the earth that nourishes us all to pursue these things?

These thoughts were disturbed by a grunting from beyond the stream. My patrol leader stood up to see what was happening. He laughed, and I jumped up to see two white rhinos mating. They were grunting, squealing and embracing one another, the male's ears twitching back and forth.

I stood watching them, laughing too because it was comical, yet also awed to be witnessing the sacred moment of a new, wild life being brought into being. Life goes on, I thought as I watched them, realising suddenly what a miracle was captured by those three simple words. *Life goes on* ... It is broken, time and again, by those who would destroy it for their own gain. And yet it goes on. I felt the power of this ongoingness trembling through the earth, radiating from the ponderous, clumsy rhinos to me, to the trees, to the hills all around. I could feel my own new birth stirring within myself, the birth of some new wisdom and awareness. An awareness of connection, of ongoingness; an awareness that here, intertwined, were perhaps the two most sacred strands of life.

———•———

Behind our compound at Mduba was a koppie created by massive boulders of sedimentary rock, some as high as six or seven metres. It was home to many creatures – venturing up there, I came across dassies basking on top of the rocks, skins that snakes had shed, bulky rock monitors clambering over the boulders like relics of the dinosaur age, porcupine quills under the rocks. Once, I came across a leopard spoor – I could even see the faded claw marks on the rock, created as it took a giant leap to the top of the boulder. I felt so humbled to be sharing this space with other creatures.

The top of the koppie offered a view over sweeping plains

covered by red grass and patches of bare red soil. Giant umbrella thorns towered over the grass, and often, especially at sunset, the plains would be teeming with impala and giraffe.

I remembered the many evenings I had spent on the boulder behind my homestead in Hlabisa, and was drawn to repeat the ritual here. As sunset approached, I would feel an invitation tugging me, and would go up to spend three or four hours sitting on top of the sun-warmed boulders. It was safe and provided me with precious time for solitude.

I would wait patiently for the night to swallow me, watching as the sun sank to hug the horizon, before vanishing slowly until the last golden gleam faded on the rolling hills. As the light died, the zebras would start braying and a spotted eagle-owl would start hooting while the darkness enveloped me. Bats darted in the sky, sometimes passing so close that I could feel the soft wind from their parachute-like wings. I remembered the song we sang in my village when we were young. It honoured the smoothness and the lightness of their flight, and called them the Kings of Darkness (Lulwane, Mlulwane Lizinkosi zomhlalabusuku). We would throw something in the air, and watch as the bats came gliding down to snap at it, marvelling at their speed and accuracy.

I remember coming up the koppie one evening, lying back to contemplate the stars as they came out one by one in the moonless sky. I tried counting as they emerged but soon there were too many. From far off I heard the sound of lions roaring. Soon I was in utter darkness, and could not make out the trees that were two metres away. I sat watching shooting stars on the horizon, listening as the creeping, unknown sounds penetrated my consciousness.

The spirits of the night floated like dry grass on the water. There was a rustle in the grass, setting my heart racing as I visualised a leopard sneaking up from behind, or an elephant creeping up. Shivers embraced me as a cool evening breeze stirred; I could feel the hard contours of the rock digging into my buttocks. I pulled my trusted old coat close around my shoulders and allowed myself to be submerged by the wisdom of the night. Gradually, my mind calmed down, my imagination stilled, and I could give myself to the present. I continued with my solitary meditation, contemplating the experience of being in the heart of darkness in pure wilderness, alone under the African sky.

In time these evening visits became a ritual, echoing the one I had at home. But I only went up if I received a clear invitation from the dassies, snakes and porcupines. I needed to respect the place and the other creatures sharing it to sustain a smooth and clean relationship with the land, and to enable me to keep my spirits receptive to what might come.

One night, as my eyes were adjusting to the darkness, I heard something shuffling through the sand. My headtorch revealed two figures with long ears and pig-like snouts – aardvarks. I was surprised to see them, for these nocturnal creatures are shy. I watched them for a few seconds in the beam of my torch, then switched it off. For me it was enough to feel their presence, and I didn't want to disturb their activities just to satisfy my curiosity. I remembered the stories my dad had told me about these animals, how they symbolise good luck; about their fearlessness in fighting with termite soldiers when raiding termite nests. I sat in the darkness, listening to the aardvarks turning the soil. I could hear their claws scratching against the rocky surface, the movement of their thick tails as they brushed the soil while they dug. A spotted eagle-owl hooted from a nearby tree. The sound floating across the hills, so clear and so loud, brought a cold shiver through my body, for in my culture, owls are omens portending death.

It felt as if the bats, aardvarks and owls were messengers, emerging from the darkness like night travellers, scholars from the East. I knew they were bringing me messages of wisdom and courage to follow my true and powerful path. I knew too that the power of that path lay in humility. By remaining humble and empty, I could create a space that enabled me to pay homage to these messages from the Great Mother Earth, and from the divine – for everything on earth reflects the soul of God.

When the aardvarks left, I crawled on my hands and knees to where they'd been digging. The fresh scent of the newly turned soil greeted me as I placed my hands on the scar where the soil had been dug and paid homage to the termites that had lost their lives to give life to the aardvarks. I thanked all the sounds and beings of the night, for allowing me this sacred means of communing with unknown worlds.

I have continued this ritual of meditating in solitude and darkness throughout my life. Each time I perform it, I seem to go deeper and deeper. Sitting in darkness is a powerful incubator for the transformation of thoughts, body and soul, for all life stages include light and darkness; the ritual has enabled me to forge a deep connection with the richness of life, for all living organisms begin their lives in the darkness of the womb and all creation emerged from the darkness of empty space. It has become a doorway, enabling me to step from the material world into the sacred, invisible world. Sometimes it is an orchestra of musical instruments, with a chorus of inner voices harmonising in my heart. Sometimes it is a safe harbour, where I can escape the rough seas of a demanding and tiring world, or a pillar that I may lean on to find the strength to face disappointments, defeat and despair. It has given me sufficient courage to follow my own path, the brave heart I need to face down my fear of making mistakes.

People go to great lengths to escape the pressures brought on by the modern world, often turning to drugs or alcohol. But these are only temporary medicines, for the new dawn will greet them with the same chronic illnesses of depression and anxiety. People chase after luxury, believing it will bring happiness. I was so fortunate to have discovered at the age of twenty that by sitting in the lap of the earth under the African stars in the pitch dark with nothing but my own body and soul, I could hear the music of the sacred world and set my spirit dancing.

———•———

Six months after I started volunteering, I moved to Stezi compound. It was on top of a small hill wooded with the dark graceful trunks of the wild syringa. In this compound, we stayed in separate small rondavels. I'd been worried about leaving Mduba because Dumisane had been like a father to me, but luckily one of the rangers at Stezi was Nkalakatha Leonard Nxumalo.

Nkalakatha had a long connection to my family – his father had raised my mother, and Nkalakatha had grown up believing my mother to be his biological sister. My mother was always invited to the ancestral ceremonies his family conducted, and Nkalakatha and his father would come to help with our ceremonies. When I came to Stezi, he said, 'Well, as you are coming from my sister's hand, I must take care of you.'

Nkalakatha is a strong man with big hands and a deep, sonorous voice. He is a phenomenal tracker with a powerful knowledge of the bush, and is extremely brave. I've seen him fighting with a crocodile that nearly took us as we crossed a river, and chasing a lioness that we had run into – yet he's also very respectful of wild creatures. He only did that to save our lives. Once when we were together he succeeded in arresting four armed poachers, even though we only had one rifle between us.

The other guide there was Baba Mhlongo, an old ranger who could interpret the cries of animals – he could tell if they were calls of alarm from being under attack, or if they were sick, or if they were just calling to each other. Between them they were wizards at reading the narratives of animals that you could not even see, one through the animals' tracks and the other through the animals' calls.

Although I was based at Stezi, I moved around a lot. The park was short-staffed, and I used to fill in wherever there was a gap. This meant that I often went back to spend a few days at Mduba, and one of the activities I got involved with there was rhino monitoring and notching. All the black rhinos had been darted to have their ears notched. When we did rhino monitoring, we had to get close to the animals to study and record the pattern of notches so that we could identify them accurately and keep track of their movements. This was necessary because, although rhino poaching was still at a low level, the black rhinos were already endangered.

Baba Thabethe led the rhino monitoring. He liked my passion and curiosity, and was keen to get me involved in this work. But my first day of rhino monitoring did not quite go as I imagined it would …

We set off on a path that had been washed clean of animal tracks and footprints by the rains. At my feet, the telecoprid dung beetles busy rolling their balls were being threatened by the kleptocoprids, a type of dung beetle that steals the balls of other dung beetles instead of creating its own. I paused to watch a deadly battle between these creatures – I could hear the cracking of their carapaces as they wrestled on the ground. I was cheering on the telecoprid, the rightful owner of the dung. 'Get that lazy bastard!' I yelled. But the kleptocoprid took the ball, leaving the telecoprid looking confused and forlorn.

We walked on, heading towards the pan. In the days after the rains, the air seemed cleaner, the sky bluer, and all the animals were celebrating. Impalas were pronking and rutting on the ridge. Bee-eaters swooped through the air in bright flashes of colour, snatching insects with deadly accuracy.

As we walked through an umthombothi forest, the smell of fresh black rhino dung hit me sharply. I felt a kick of adrenaline, as I looked at his footprint freshly pressed in the soil, the nearby urine and dung still warm. We had come to find these animals, but the awareness that they were close by still came as a shock. Baba

Thabethe turned down the volume of our two-way radio. I crept behind him with one thought hammering in my head: get behind a tree!

Red-billed oxpeckers called nearby, then fell silent. It seemed as if everything was watching, waiting for us to collide with the creature. Sweat trickled down my face; our footsteps seemed to grow louder and louder, my breathing roaring in my ears as we walked.

Suddenly, a black rhino burst from the bushes about forty metres away and came charging towards us. He was heralded by a great cloud of dust, and swarms of the black flies that infest the wounds from parasites often found on these animals. The furiously buzzing flies seemed to be goading the rhino as he accelerated towards us. As I turned to run, my foot caught on a tussock of grass and I felt myself crashing to earth.

The terrifying sound of snorting and grunting filled the forest. I had no time to get up and find cover. I just had to lie, frozen, on the ground, waiting for the horn to perforate my body. His footsteps drew closer ... seven metres, four metres, two metres ... I closed my eyes. Then he was right upon me. I heard Baba Thabethe's petrified voice whispering to me to keep lying down. The rhino was so close that I could smell his breath and when I half-opened one eye I could see the ticks on his skin.

The flies flew from his body and started biting my naked legs, but I forced myself to keep still. I thought of the sharpness of his horns, of the thin edge between life and death. I willed the earth to envelop me, to keep me safe and hidden. The rhino kept on standing next to me, staring at the ground, while my heart drummed loudly in my chest. Time seemed to freeze. Baba Thabethe shouted and threw some branches at the rhino, until at last, mercifully, he ran away. I sat up and dusted off my clothes.

'Bekushubile Mfana wami' (That was tough, my boy), he said, helping me up and passing his water bottle to me.

I knew the black rhino was the most feared and respected

creature in the game reserve and I was drowning in fear. With only one kick, that black rhino could have crushed my skull. With one thrust of his horn, he could have shattered my spine. Tears streamed down my face, cold penetrated my veins.

I took a sip from the water bottle. I was trembling all over and could hardly stand – I needed my colleagues to support me. When I'd recovered enough to walk, we followed the path that was now winding through a thick bed of reeds. Holding my knobkerrie in my hands, I shouted, 'Sweet Jesus, are we walking through the reeds after this experience?'

The path took us straight into the depths of the reeds. I could smell the fresh scent of buffalo dung. I could smell danger and waited for another terrifying encounter, this time with the buffaloes. A bushbuck barked and bolted at the sound of the reeds being moved – we could tell it was running away from us. At last the path regurgitated us on the other side of the reeds, onto the bank of a broad pan. We stepped out, relieved to be free of the reeds, welcoming the sweet scent of wild basil.

We sat on the trunk of a sycamore fig tree that was leaning out over the water. A white blanket of mist hovered over the surface. Weavers flitted about, building their nests on the branches dipping low over the water. The reflection of another sycamore fig hung in the still water on the far side of the pan.

A breeze came up, rippling the surface and setting the water dancing; the leaves of the trees joined the dance, while tilapias swam beneath the dancing waters in flashes of silver, watched by a pied kingfisher perched on a dead branch of a splendid thorn.

The serenity of this scene became medicine for my trauma, as my terror slowly faded into the landscape. I was alive and unhurt ... my heart was still beating, my breath still flowed, my limbs still moved to my will. I thought about my utter vulnerability at that moment on the ground, how this contrasted sharply with the male ego I was raised to believe in, with its delusion of power and invincibility.

Understanding my vulnerability in that moment was terrifying, but I realised it could also be a gateway to a spiritual path.

I knew that I had lost something in the encounter. I had lost my belief in my invincibility as a man, my faith in my male ego. But I also sensed that losing this had created the space for something new. I had wept with fear and found myself trembling on the edge of life, confronted by that fragile, unknown barrier that stands between the living and the dead. My vulnerability had brought me to that fearful threshold, but it had also brought me to another way of being that could comprehend the vivid, startling beauty of life, of all living things and all stages of life – including dying. And so my vulnerability was both a threshold and an incubator, like the pupa in which a caterpillar becomes a butterfly, and it gave breath to the winds of change blowing through my heart.

After this incident, I was terrified of black rhino, but Baba Thabethe encouraged me to push through the fear. He knew everything there was to know about tracking rhino, how to read the wind so that you stayed downwind of them, what to do if there were no trees for cover. He taught me to overcome my fear by getting to know the animal as well as myself, by being calm, respectful and vigilant. After several operations with him, I too became very competent in the rhino monitoring process, and would sometimes be the one to instruct new rangers.

●————————●

Late one night a loud knock sounded on the door of my hut. I sat up sleepily, and looked through the bare window – I could see the glow of the moon behind the ridge. I cursed the moon for rising. A full moon draws the poachers, so now I would have to join the night patrol. With eyes still heavy with sleep, I stumbled around, gathering my torch, knobkerrie and balaclava. I could hear the footsteps of my colleagues outside, the crackling of the two-way radios, the click of their firearms as they went through their routine safety procedures.

I picked up my old coat, which looked like a relic from the Second World War – it was made of dry cotton and was heavy and itchy – and walked outside. The patrol leader gave us instructions and explained the aim of the night patrol, as the moon slowly rose over the lip of the ridge. The night was loud with the sounds of nocturnal creatures, and dark shadows that made everything look big and dangerous. The rocks resembled animals, the trees took on the shape of elephants, and I felt a shiver of fear.

The path we followed wound through the black monkey thorn forest. The going was clear under the branches, as the trees had grown tall, and bright moonlight flooded through the leaves. We arrived at our observation point and settled down. I took a deep breath and sat on the ground, gingerly lowering my butt onto grass that was wet from dew. No smoking was allowed, and no talking – only whispering if we had to communicate.

I huddled into my Second World War coat, welcoming the warmth it brought to my shoulders, and listened to the chorus of nocturnal animals. The clouds and moon wrestled for supremacy in the sky – sometimes the clouds swallowed the moon and we found ourselves in deep darkness; sometimes the moon dominated, and flooded the landscape with light.

The wrestling of light and darkness reflected my inner conflict. I was going through a tough time. My father had cashed in his pension, but instead of using it for my studies, he had abandoned our family and disappeared to spend it on women and alcohol. We were devastated as a family, and my chances of going to university had been destroyed.

I'd been crushed by the news. But as I watched the moon moving behind the clouds, I recalled uMkhulu's words during my initiation, when he spoke of how our lives have both darkness and light, how even on the blackest night the stars still shine. I had been experiencing grief, shame, anger and hopelessness, feelings that could have destroyed my soul. But when the clouds stepped aside and let the moon bathe us with its light, I felt a stirring of hope. Much as there was darkness in my life right now, there was also so much light, not least this opportunity to experience the moonlight in the wilderness. I also had the support of so many inspiring people – Sabelo Msweli had been like a true father to me through this difficult time, saving me from sinking into hopelessness, and Dumisane, Nkalakatha, Baba Thabethe and others were going out of their way to help me.

This moonlight seemed to be inviting me to lay down my sadness by reminding me of the light in my life. As the moonlight quietened my inner turmoil I came more fully into the moment, and let the experience of being embraced by these two powerful contrasts activate my senses. When the clouds swallowed the moon, I relied on my senses of smell, hearing and touch – I could hear the sound of grass being uprooted by a nearby grazer; I could pick up the scent of an elephant.

A leopard called just behind us – so close that all the details of its sound were clear. Then came the whistle of the mountain reedbuck, and the shouting of baboons far up the kloof echoed in the forest. The wilderness orchestra was performing! And with the ears of my heart I was listening, and with the eyes of my soul I was a spectator. The moment was infused with liveliness. All was bathed in the beauty of the moonlight's shadows through the branches of the tree, the moving black patches of clouds on the ground ... the muted colours and movement of the grass in the moonlight soothed my feelings of hurt and betrayal, and invited me to lay down my grief and step into the intense beauty of the here and now.

I felt myself travelling through time, taken back thousands of years by the night sounds and the moonlight to the Palaeolithic age, when humans understood nature with no devices, no scientific instruments or cameras or laboratories, only our senses. Our human ancestors used the sun to track the hours, the stars to give direction at night. They lived off the land, on the land, with the land; danced with wild animals in waking dreams and rituals, communed with the living spirits of the rocks and trees and water-falls. To be able to relive something of their experience made my soul sing with happiness. The ancient memories of that time were embedded in my genes. What I was discovering about nature was not new – it was a deep, internalised knowledge. I was not learning it but rekindling it, blowing on the embers of a flame that was all but extinguished, once more to experience its light.

Perhaps it was just for a night, but in that moment I was living as our ancestors lived. Their lives were singing through me, and I plunged into the sacredness of it.

We sat in our own stillness and silence as the fading sounds of nocturnal creatures and emerging sounds of diurnal creatures formed two powerful symphonies, competing until the morning creatures dominated. The sunlight crept over the horizon, the fresh morning air stirred the leaves and the grass. Few experiences are as moving as witnessing the breaking of dawn in the wilderness, a time when the creatures give thanks for the coming light and warmth, for surviving the hunting hours, for the food received by those that have successfully hunted. As I watched the golden light spread, I felt a further easing of my spirits. I knew I had many hard-ships ahead. But who is not lifted by the optimism of the dawn?

We walked back in silence under the rising sun. A herd of impalas watched us curiously as we walked past, but did not run away. I felt emotionally revived and grounded. I knew that, as much as I hungered to go to university, what I was discovering in nature cannot be found in a textbook or lecture hall. It cannot be shared with the leaders of nations or CEOs of corporations. It cannot be bought or sold, making it invisible to our market-driven world, where humans are reduced to consumers and everything else to a commodity.

We followed the path zigzagging along the Black iMfolozi. A fresh mist drifted above the floodplain; the air was filled with the scents of wild lavender and wild basil. The colours of the water shifted and gleamed. We stopped for a break, looking across the river at a herd of buffaloes lying peacefully, their horns emerging through the mist.

And beyond, the infinite serenity of a stately giraffe, standing still.

———————

I woke early with the sunlight trying to penetrate my eyelids. I was still staying temporarily back at Mduba, and my steel bed squeaked like a grumpy old friend as I sat up and looked out through the window with no curtain at the glittering dew on the grass. As I boiled water for coffee, I wondered which activities this new day would bring. I always asked my colleagues not to share the work programme with me. I was happy to embrace whatever was put on my table, but I didn't want to be caged by expectations. I didn't like going to bed with thoughts of instructions for the next day; I wanted to keep my mind uncluttered to allow for reflection and wisdom.

I took my book and my journal and sat outside, warming my hands on my cup of coffee. I tried to keep negative thoughts from my mind, but there were so many problems banging on my door. I'd been volunteering for well over a year, still with no sign of a job. My contemporaries who had been able to go to university were preparing to do their practical year – it was hard not to feel left behind. I'd not only been let down by my father, but had also lost out on a bursary because of corruption among Department of Education employees: there were only three of us from the district whose matric results were good enough to qualify for this govern-ment grant, yet none of us had got it. Instead, the officials had awarded it to their relatives, who hadn't obtained the necessary grades. And my father was still absent – the only news we had of him was if someone would tell us, oh, we saw him in the tavern in town and he was drunk, or he was with his new girlfriend.

I was also facing difficulties with some of my colleagues, who looked down on me because I was not earning a salary. They would dismiss things I said, or tell me that even if my ideas were good I should not think highly of myself. At first they also looked down on me because I did not have a uniform. Later, the section ranger gave me a uniform so that the other rangers could easily identify me when we did operations with poachers. But these colleagues said, well, now you have the uniform, but you don't have the epaulettes, so you are still not worth anything.

I could not understand this need to feel superior, to judge a person by something as trivial as an epaulette, rather than by what they could contribute. I felt like an outsider, a bird with different feathers. Nkalakatha Nxumalo encouraged me to ignore this, saying I should not be disturbed if other people could not understand my worth. What mattered was that I understood my life's purpose. He told me that his mother had been abandoned by his father, and yet he had overcome the hardships that had resulted. He said, 'Today, I'm old, I'm fine, I'm working … these are the hardships you have to confront, and I'm here to make sure you don't go to bed hungry, or go on foot patrols with no breakfast. Don't hate your father. Learn from his mistakes, and go forward.'

I knew that feelings of inadequacy had created many wounds in my heart. All of us in my village grew up with feelings of not being good enough, because we would sometimes go hungry, because we were so aware of scarcity in our households; we walked the long road to school on bare feet, we were ashamed if our school uniforms were old, or torn, or too small. We lived in a world where status was measured by material wealth, or power or position, yet it was almost impossible for us to acquire these things. We did not realise that all the wealth and power in the world can still leave you feeling empty. It was these feelings of inadequacy that drove some of my peers to become hitmen in the cities, or to turn to alcohol or drugs. These feelings also created divisions in our communities, leading to jealousy and resentment.

I tried to pull my mind away from these worries by focusing on the weavers building their nests in the nearby umbrella thorn. Some of the lazy males were stealing grass from the other nests when their owners were absent – just like those corrupt Department of Education officials, I thought, stealing the bursary I had earned through my hard work and robbing me of an education.

But as I watched the industrious little creatures, I began to lose my feelings of anger and distress. I remembered once seeing a black mamba creeping through the branches of a sausage tree growing at the river, from which many weaver nests hung over the water. I could hear the chicks twittering inside the nests, and was sure the mamba was after them. But as the snake got near, all the weavers flying about came together and started jabbing at it with their sharp beaks. They flew in from all directions, always attacking from behind to avoid being bitten. At last the mamba lost its grip on the branches and fell into the water below.

So, even if they sometimes steal from one another, these birds can also work together and create a powerful force when they are united. I remembered my first battle for nature, inspired by the destruction of an umbrella thorn much like the one the birds were nesting in. The collective protest of us children, small as we were, made the adults take action to protect trees in the future. None of us can live without the support of others – how much better the world would be if we could all recognise this and stop crushing other people in our race to 'the top'.

My Mbatha clan has a special relationship with weavers. We have no totem animal, but our izithakazelo, our praise name, recognises us as the warriors of weavers – nina abantaka eyahlala emhlangeni wawathalala. These phrases compare us to a flock of weavers that fly down and land on the long grass stems, causing the grass to bow down under the birds' weight, as if bowing out of respect.

The praise song comes from a time when the Mbatha clan was fighting with King Shaka's clan. Our clan was defeated, and ran up a steep hill to escape. Our warriors then started rolling big boulders down onto Shaka's warriors, until they overcame their enemies. When Shaka's warriors conceded defeat, they called for the Mbatha clan to come down the mountain to negotiate. My clan was using white shields, which made the warriors resemble a flock of birds as they descended the slope. They flattened the grass when they sat down at the bottom of the slope, causing the grass stems to bow before them.

Thinking about this made me feel hopeful for the troubles in my own family. My Mbatha ancestors had faced difficulties and overcome them by working together. We would overcome this problem, somehow. My father would come home, seeking forgiveness. Perhaps we could do rituals to enable the ancestors to help us reconcile.

I gave thanks for the people at the reserve, like Dumisane, Nkalakatha, Baba Thabethe and Sabelo Msweli, who were helping me to weave a strong, safe place in the world by giving me support, guidance and hope. I recalled Nkalakatha's words that I must believe in my own journey. And I knew that each day I spent in the iMfolozi was strengthening my belief in this journey. I became a volunteer because I wanted to gain the knowledge needed by a game ranger, and to make myself eligible for any available job in the game reserve. I was hungry for knowledge and opportunities. But I was learning that I had many other hungers and thirsts, and my volunteer work was serving as food and fresh water to satisfy these needs. I was learning how to live mindfully and meaningfully, I was growing in my mind and soul, and pleased that I was developing what I needed for this inner journey. Nkalakatha was

Learning about resilience from the weavers

right: I did not need those guys who worried about epaulettes to understand the gains I was making. What mattered was that I understood them.

The support of Nkalakatha, Dumisane and the others helped me conquer my feelings of inadequacy. But I was also discovering how being connected to nature brings self-awareness and self-acknowledgement. Connecting with nature doesn't boost your ego in the way that flattery and possessions do; instead, it helps you to step away from your ego by helping you to dissolve your separation from other living things. It connects you to the glorious dance of life, and helps you to recognise the inner beauty and worth that come from being part of this dance. This brings a feeling of 'enoughness' and wholeness that no possessions or prizes could ever bring.

My coffee cup was empty, and my colleagues were calling me to join the patrol. I stood up and bowed my head in homage to the weavers for all they had taught me, but they were too busy weaving to notice. They were just busy being weavers, and did not crave applause.

One autumn afternoon, while I was still temporarily located at Mduba, I was walking on patrol with Dumisane and Baba Thabethe through a patch of dry turpentine grass. The strong, oily scent that gives this grass its name stung my nostrils, making me sneeze. The trees had started shedding their leaves, and held naked limbs up to the grey sky. The soil was bare of vegetation, revealing patches of sand of different colours. A small whirlwind stirred the dust, swirling it into a spiral up to the sky. In isiZulu we call these isikhwishikazane – they are too small to cause much damage, but there are beliefs that the spiral shape identifies the whirlwind as an evil spirit. Nearby, a lonely wildebeest bull was jumping and snorting despondently. A grey miasma wrapped the hills, making the land look blurred, and a strange sense of sadness clung to the land like an evil omen. An emerald-spotted wood dove lamented in the woods nearby.

We sat down for a break. I opened my water bottle, glancing at my old canvas boots that were covered with dust. My legs were itchy from tick bites, and full of small, bleeding scratches from the thorny bushes we had walked through. I gulped the tepid water, trying to ignore its muddy taste. The surroundings were blanketed by stillness. Nothing moved but the lonely wildebeest across the valley, now drifting forlornly across the red soil like a wrinkled old balloon scudding along in the wind.

Everything seemed weighted by sorrow: the dry pan, cracked like a frowning forehead; the terrapin scuffling the dry clay, trying to find some mud to bury itself in; the bees flying around the pan and sucking what moisture they could find through the cracks.

I walked away from my comrades to find a tree giving shade. As I crossed the cracking mud, I noticed small droplets of blood. I knelt to examine them. They looked fresh; when I touched them, I realised they were still wet. How had this blood come to be here?

What animal had shed it? Was it injured? What had injured it? I knew a skilled tracker would be able to read these details.

Drawing on my still-limited tracking skills, I followed the wavering trail of blood. It led me to a forest of wild camphor bush. As I approached, the dove that had been calling flew up. I heard the sounds of my comrades' careful footsteps behind me – they had come to see what I was tracking, and were also following the trail, their eyes fixed to the ground.

The harsh croaking of ravens penetrated the silence. Baba Thabethe bemoaned the sound, for in our Zulu culture ravens symbolise death or a lost soul, and here we were following a trail of blood. The English collective noun for ravens, an 'unkindness', seemed very fitting at this time. Somehow we lost sight of the trail, but we could see where the animal was staggering, and the way the grass had been flattened by its weight indicated the direction in which the animal had walked.

In the depth of the wild camphor bush, the smell of death penetrated us all, while the unkindness of ravens moved restlessly around the branches of a nearby tree. We stopped to listen and sniffed the air to catch the scent of animal gut – it seemed fresh. We sat on the ground under the trees to get a better sense of this animal we were tracking. The trees wept old, dry leaves; I could hear their soft whisper as they landed on the grass or on my head. I gestured to my leader to suggest that we continue with the patrol, for there seemed to be nothing that we could find.

Then we heard something bleating in the heart of the bush. A newborn animal: had we been tracking a creature that was about to give birth? We waited for the mother to comfort it, but the bleating continued, the calf's poignant, innocent cry floating through the silent bush. We sat wrestling with our thoughts and our choices. Do we help it? Or do we leave everything in the hands of nature, as we have been trained to do? At length we decided to see why the mother was not comforting her young one.

The mother was easier to track now as we could trace her direction from the sound of the bleating. As we pursued the sound, a trembling gold fluffy creature came through the bush towards us – a newborn kudu. Its shiny black eyes held the beautiful innocence of a baby's face. We stopped as it approached us on shaky legs. A few flies had started feeding from the moisture on its skin – it was still wet with birth fluids, and we could tell that the mother had not yet licked it. Its big ears were lowered, hanging on the sides of its face like old stocking caps. Why was this little one alone? And where was the mother?

We continued in the direction the young kudu had come from. Then we saw her – she was lying on the ground, her tongue sticking out, with no signs of life. The clearing was filled with the heavy buzz of green flies, drawn by the scent of the placenta. I gently touched one eye – it did not blink, and small ants had started crawling around her nostrils and eyes.

The calf was quiet now, but the ravens kept up their sinister croaking. The mournful sound of the emerald-spotted wood dove floated through the thickness of the camphor forest. I was puzzled by the dead kudu – was it hurt? I studied the body, but could find no signs of injury. Water was plentiful by the river ... what could have killed her?

I grieved for the calf, knowing it had little chance of survival. My experiences in the wilderness had helped me to open the door to my emotions, and I sat down and wept. I felt a deep resonance, for my grandmother had died giving birth to my father. The calf trembled against me as I touched the fluffy head, and blinked up at me with soft black eyes. Its new fresh breath touched my nose. We were exchanging breaths, and it felt like prayer. Then, with a heavy heart, I stumbled away, for I was not allowed to intervene.

As rangers, we may intervene if the animal is an endangered species. It was heartbreaking to walk away from a creature that had approached us with such vulnerability and trust. It would not be long before the scent of blood drew predators and the calf became a meal that might feed other hungry young. We walked home in silence, each lost in our own thoughts and regrets. The grey smoke of the evening haze covered the hills; the setting sun hung lonely over the African savanna. I thought of sunset and sunrise, of death and new life and new hope. Whatever the day had showed me was mirroring the grief, loss and new birth in my own life. I knew I had to accept and embrace whatever pain grief brought, as part of myself. Cruel as it was to have had to leave the calf, it was a tough reminder that we cannot always 'make things better'. Sometimes we have to accept sadness and grief, and allow time to heal us.

————•————

Back sat Stezi, I found myself on foot patrol with Nkalakatha and Baba Mhlongo. These two rangers understood me well, and could see that I was keen to absorb everything I experienced in nature. They supported me by going even deeper into the iMfolozi wilderness. We crept through terrifying thickets, traversed sweeping valleys. Sometimes we had to run to save our lives, sometimes we had to stand still – and I learnt which action to take for different threats. Each day brought new ingredients of life to embrace, and new wisdom and skills to master.

On this day we were walking through a dry winter landscape. The grass was drying out, the trees were completely bare of leaves, and all the paths in the bush were exposed by the thin vegetation. Dry leaves on the ground crackled beneath our feet, like potato crisps. The animals were starting to congregate at the rivers, as the mud holes were drying out. We walked past warthogs bathing themselves in mud that was almost dry. The winter wind set the abandoned nests of the southern masked weavers dancing so vigorously that some fell to the ground.

Gleaming rays of winter sunlight pierced the thickness of the bush, and slowly the grey mists retreated. I used my sunhat to swat clouds of flies away from my face. As we headed towards the river, we could tell it was busy from the splashes and the bellows of the buffaloes. Along the river we followed a line of stately old fig trees. A troop of baboons played peacefully among the trees, chattering and swinging from the ends of the branches. Shaggy-haired nyalas wandered through the reeds, nibbling green leaves on the trees near the river, their dark forms outlined against the silver glow of the sun on the water. The spiral horns and delicate tawny legs of a drinking nyala bull were reflected in the still surface.

Across the river, impalas were pronking with the joy of the day, while buffaloes grazed peacefully on the short, green grass near the riverbank. I took in this scene, painting a picture of what I was observing with my heart. I did not own a camera but I was learning that taking a picture with your heart can be more valuable than a photograph – sometimes, looking through the lens creates a barrier between you and what you are observing.

I felt a smile in my soul. I let myself sink into the sounds: the high, wistful call of the African fish eagle, the grunting of the rhinos, the brassy trumpeting of elephants. I laughed at the young fluffy buffalo calves as they ran playfully, tossing their heads from side to side, at the old buffalo bulls sparring on the riverbanks.

As we moved on with the patrol, we were alarmed by the shouting of the baboons. Drawing on what Baba Mhlongo had taught me about animal language, I interpreted their shouts and called to the others, 'Predators around here.' Within fifteen minutes, we came across a pride of lions hunting a grazing herd of buffaloes. One lioness was leading the others, as they formed a V shape and stealthily approached the buffaloes – a well-known hunting strategy.

We sat down to watch. Lions often hunt at night, and it was rare to witness a kill in daylight. This was my first experience of witnessing a lion hunt, and I was curious but also reluctant to see the death of a poor buffalo. The buffaloes came out of the bush, walking across the open grass to reach the water beyond. The lions were now completely separated from one another, with one big male waiting alone down by the river and three females hidden in the small bushes nearby, still holding their V formation.

As the buffaloes started drinking, the big male lion charged. His long black mane rippled in the wind; his blood-curdling snarls flew across the grass. Startled and panicked, the buffaloes ran wildly in all directions. There was a cacophony of calves bellowing, water splashing, the clatter of hooves as they slipped and scrambled on the rocky riverbank. A lonely elephant bull nearby trumpeted in alarm, and swivelled his trunk; the place was enveloped in the scent of dust and fear. The red soil and dung disturbed by the stampeding animals swirled in clouds behind the herd as they galloped frantically towards the thicket.

As the dust died down, we realised that the lions had caught one big buffalo – we had missed it in all the commotion. All four lions were sniffing the dead buffalo as if saying a prayer before eating. The male started eating, and the females soon followed. Drawn by the scent of the buffalo gut, vultures appeared high above us, circling in a descending spiral. Nkalakatha pointed out the lappet-faced vulture, the buka izwangomoya, as it landed on a sycamore fig. The soft mellow whooping of hyenas started up from a nearby stream.

We could hear the crack of bones as the lions fed on the carcass. As the lions satisfied their appetites and started to lose interest in eating, a bateleur flew down and began tearing off small pieces of meat. The lions watched it eat, their bellies bloated, their faces painted red like those of warriors going to battle. They drifted away to lie in the shade nearby, their long, red tongues lapping the air as they panted.

Led by Nkalakatha, we walked cautiously towards the carcass, under their gaze. They were too lazy and too full to move, but they growled so ferociously that my stomach turned to water. I nervously watched their neat, rounded ears flicking, their tails twitching. But Nkalakatha continued walking calmly straight to their kill, and cut off a chunk of meat big enough for the three of us. By now the lions were all standing, watching us closely, but as we quietly moved away they collapsed back onto the ground. The male lion approached the carcass, sniffed it, licked the place where Nkalakatha had cut the meat, then laid his head over it. The tree nearby had filled with hundreds of vultures, jostling for space on the branches. They waited patiently, knowing that waiting would reward them with a better share of the kill. Waiting is a critical skill in nature. It teaches you to share, and to respect and embrace fully what you have been patiently waiting for.

Fresh blood from the meat dripped down on my back as we walked back to camp, and dried on my torn khaki shirt. Like the lions, I was stained with blood, the blood that connected me, too, to the buffalo whose life had been sacrificed. I was grateful to the buffalo for the food of the day, for food had been scarce for some weeks. The loss of the buffalo's life had given food to many – the lions, the bateleur, the vultures, me, my companions, and countless other organisms. What I was carrying was not just any chunk of buffalo meat. It had been given, not only by the buffalo but also by the lions. Consuming this meat was thus an act of powerful communion, granted to us that day by Great Mother Earth.

I was so moved by the generosity with which the lions had shared their food. I thought of the children I had seen scavenging from dirty dustbins in the towns near my home, the food waste in overflowing bins outside supermarkets … How much could we learn from the lions, that take only what they need and leave the rest for others?

I looked at the backs of my companions going before me, and let my sadness melt back into the ground under my feet. These two had never hesitated to feed me when I was hungry. Like the lions, they were always willing to share what they had. There is much greed in the human world, but also much generosity.

This is what will save us from ourselves.

———

In my third year of volunteering, I moved again – first to Zimambene (the place of mambas), where I worked temporarily as a security guard at the craft shop and information centre. After months of working without pay, I was grateful for this employment, even just for a couple of months. It was organised by Erica Robertson, the wife of one of the section rangers, Dave – both of them had been very kind and supportive. They sometimes bought me food in town, although I struggled to eat it because it was too 'white' – very different from what I was used to. Dave was inspired by my passion for wildlife, and would find all the opportunities he could to give me new experiences. Being employed meant that for those few months I could take care of myself, and did not rely on the generosity of others for my food.

After Zimambene, I was moved to Masinda, a secluded camp that faced the Intshevu Stream. This is a beautiful, thickly wooded stream that few are brave enough to venture along, for its thickets conceal fearsome creatures that can ambush you and its waters often harbour crocodiles. There were many older rangers at this camp who knew my father and had known me as a young boy. They treated me as their own son, and I tried to repay their kindness by helping them with reading and writing, as many were illiterate and could only sign documents with a thumbprint.

The rangers I worked with most often were Sonto Mthembu, Sbusiso Zulu and Mandla Gumede. Mandla was closest to me in age. Sonto was a leader in our group, a strong Tsonga man whose fit, muscled body gave him the nickname uMagwalakaqa (the toughest one). Looking at him you could imagine him wrestling elephants, but he was a gentle spirit with a kind heart.

It was Sonto who taught me a different way of doing black rhino monitoring. Baba Thabethe and the other Zulu guides, infused with the macho culture of the Zulu, would tell you to stand behind a tree and call the rhino to you. Then you would have to keep moving around the tree to make sure it stayed between you and the angry rhino, while at the same time trying to inspect its ears for notches. If you stumbled, you'd be finished. Sonto would get us all up into the tree before the rhino came, then call to it with a high wail that mimicked the cry of a baby rhino. From our high perch we could easily (and safely) check the notches on their ears.

After two years of volunteering, I was developing a unique relationship with wild animals. I was not a biologist or zoologist, but I was engaged in deep research and gaining deep knowledge. I began to know the animals intimately, identifying individual animals by their unique characteristics – a white rhino with a cracked horn, a lion with a scar on the left side of its face, an elephant with one short tusk. When you spend enough time with them, you start to recognise these distinguishing marks.

But I was getting to know the animals not only through my senses of smell, sight and hearing, but also through an inner sense, the sense of my heart. I was learning that most animals are hyper-receptive to messages through telepathy – and that for me to be receptive to these messages in turn I needed to cultivate an attitude of humility and calm. The most profound lessons of non-human wisdom were learnt in my heart, not in my mind. I was led by my intuition, which allowed these teachings to reveal themselves to me.

A great variety of animals used to wander through Masinda. The camp was fenced, but elephants, drawn by the smell of delicious groves of scented-pod acacia, had broken sections of the fence and become frequent visitors, along with various antelopes.

But the animals I developed the deepest relationship with were those in the troop of baboons that used to sleep behind our camp.

We reached a greater level of understanding with these creatures than with the other animals. Through our discipline of keeping food inaccessible, we taught them to respect our borders, and we respected theirs. We never had problems with them breaking in and marauding in our houses. Over time, my connection with the baboons expanded into oneness, and deepest love. It was similar to the love between a dog or horse and its owner, but also different, for these animals did not depend on me to care for them. We were sharing our space as independent, free beings. One of my most poignant insights into the depths of animal emotion was to come from these creatures.

On one long winter night, I woke up to a great cacophony of shrieking. The deep barking of the male baboons vibrated through my windows. The babies were screaming and, beneath the deafening cries of the terrified baboons, I could hear the growling of a leopard. I eventually fell into an uneasy sleep as the battle continued, my dreams disturbed by anxiety to know what the commotion was about.

In the morning, while I was sitting on the veranda with my journal, some members of the troop came past, walking slowly under the trees. I noticed one limping male and one female with a fresh wound still bleeding on her face. With one hand she was holding a dead baby, lying awkwardly as if its back were broken. The mother was stroking her baby's face, comforting it as if it were still alive.

I could see such sadness in her eyes. Sometimes she went off for brief periods to search for food, leaving the baby on the ground. While she was gone, another male took over, holding the dead baby until the mother returned. He too was stroking and grooming it as if it were alive. Watching them, I was struck by how similar they are to us in the way they mourned and the way they comforted each other. The encounter gave me a window into their pain and sorrow.

In Zulu culture, we too show love for our loved ones even after they have passed away, by respecting and taking care of their bodies. We acknowledge their souls; we consider their spirits sacred. Watching the baboons showed me that animals feel the same grief we feel when we lose the ones we love. I empathised so deeply with those baboons, for my family carried the spirits of its own lost infants – my brother who was lost in a miscarriage, my sister who passed away at only nine months.

Domestic animals often demonstrate incredible love and devotion. My brother used to work in Johannesburg, leaving his two dogs at home in our village. When he would come to visit us, he would phone to say he was coming, and straight away his two dogs would go to the bus station and stay there until he arrived. And I remembered Nkanyezi, the horse who knew my father was coming back to his rondavel and waited there to show him her foal. There are many stories of wild animals showing such wisdom and strong emotional attachments to humans and one another – dolphins, whales, elephants, even birds.

The baboons' behaviour pushed me to think deeply about the cruel treatment so often given to animals, as if they have no feelings and emotions – the elephants forced to carry tourists, the bears kept for bile farming in China ... The baboons' grief reinforced my respect for every living organism. It taught me to grant them the same respect as I would my family members, friends and relatives. It strengthened my bond with wild animals and created the deepest love. It reminded me that I am part of a web that connects all living things, and that any suffering by another creature reverberates in me. I am the pangolin that has been stolen from its habitat and traded for its meat and scales; I am the blue shark that faces the sharp blade, severing its fin for soup; I am the rhino that faces the barrel of a poacher's gun, for my horn to be sold as an aphrodisiac; I am the tree falling to the blade of a chainsaw in the Amazon rainforest, cleared for the planting of palm oil.

Our human ego pushes us to disconnection, and blocks us from embracing non-human wisdom. All animals, great and small, have much to teach us, but the world has become dominated by a culture that asserts human superiority. This disconnection does not serve us, and has created a great inner loneliness.

I understood then that one of the keys to repairing our broken relationships with other species is to start appreciating the depth of their emotions. But I was also coming to understand how repairing this relationship would heal our minds and spirits, that creating a meaningful connection between humans and other species will save not only the environment, but ourselves.

———

The inqunqulu (bateleur) circled, soaring high above us, calling and flapping its wings. The echo of its ancient, haunting call faded across the cliff. This bird seldom calls, and in traditional Zulu culture it's believed that its call is a portent of something bad. The inqunqulu is recognised as a spirit bird, reflecting both the invisible and visible worlds. I watched its flight, feeling goosebumps on my arms. My leader softly asked the bird to hush. As it flew out of sight, a lone baboon shouted across the hills.

The rangers had taught me to heed all the signs and languages of animals. Messages are everywhere in the wilderness, like road signs in the cities. What message was this bird bringing to me?

We walked on, following a trail through a stand of wild potato bush, bringing the smell of roasted potatoes to our nostrils. As a child herding goats in my village, I was told by the izinqwele that the smell of potatoes in the bush means that a big snake such as a mamba is shedding its skin. The elders spread this story, perhaps to caution us from rooting in the dense potato bushes in search of red ivorywood fruits or other delicacies – the dense foliage of the potato bush may indeed be a haven for snakes. Later, I learnt that the smell is released by this plant during photosynthesis. But I was still stuck in my old beliefs, and I could not help looking around for the snake that was causing this smell.

The path we followed was full of fresh porcupine droppings, with a few quills scattered along it. My boots were wet from the morning dew, and I could feel a tick biting me between the toes – how my heart cried for a break so that I could remove that tick! We stopped at last at a fork in the path – one way zigzagging into the wild camphor bush and the other heading straight into the river. Much relieved, I sat down and removed the tick.

From the depth of the wild camphor bush, a powerful sound swept through the forest. It sounded like a cat, but we had also come across fresh rhino dung. I jumped up to work out where it was coming from. There was silence, then another wild growl echoed through the trees. We started creeping towards the calls.

We tried to walk in silence, but the dry, fallen leaves of the camphor bush crackled loudly underfoot and our hearts were beating as heavily as African drums during the ritual season. Sweat trickled down my back and my heart pounded as my feet carried me towards this unknown creature. As we came closer to the sounds, we realised that we were stalking a leopard.

Suddenly, the calls stopped. A stillness blanketed everything, as if awaiting a sacred spirit. As we crouched and peered through the foliage, a porcupine emerged, its quills raised, ready for an attacker. Behind the porcupine was a young leopard, its ears forested with quills.

The leopard came around the porcupine and cautiously made another strike, using its paw to slash at the front of the porcupine rather than coming from behind. The rattling of the quills and the snarls of the leopard swept the bush again, alarming two hadeda ibises, which called out and flew away. The leopard was now very wary of the quills. It had failed again to overcome the porcupine, and quills were now embedded in its left paw. It limped away, and lay on the ground.

With tears in my eyes I watched as the porcupine disappeared into the thickness of the bush. In the corner of my eye I could see the unsuccessful leopard – the one that our society might call a failure. But the hungry leopard understood that persevering leads to a deeper understanding, and deeper understanding gives birth to wisdom. His hunger was pushing him, but he had shown wise judgement to retreat, as further injury from the porcupine quills could disable him and make it harder to hunt in the future.

As a young Zulu man, I had been taught to see failure as the worst thing. Failing or retreating was seen as a weakness. We were under so much pressure to be recognised as successful that fear of failure could paralyse us, or push us into bad choices. We heard of young people committing suicide because they failed matric – fear of failure brought them such darkness that they believed they no longer deserved to live. This fear can push parents into sending their kids far away to boarding school, to grow up without their families around them.

But the young leopard faced no such fear. He was driven by hunger, not by a desire to win or be seen as successful. And because he was neither blinded by the fear of failure nor suffering from pressure to succeed, he could keep his mind clear. With a clear mind, he could recognise that it was better to retreat from this opportunity to eat so that he could keep himself strong for the next one.

Watching the leopard and the porcupine, I came to see that our wisdom grows through failure. Failure is not something to be feared. It is part of life and one of life's greatest teachers. Whenever I am tempted to make foolish choices because I want to prove something, or because I fear failure, my memory gives me a little poke with a porcupine quill. Just to remind me to step away from my ego, take a deep breath and make a wise choice, like the leopard.

———

When the hard knock rang through the night, I pretended to be fast asleep – there was no plan for us to work that night. But the knocking persisted, and radios were crackling outside my hut. I forced myself up and staggered to the door. Mandla was outside, telling me they'd received a message that poachers had been heard hunting in our section.

This was before the days of serious rhino poaching, and these poachers were usually hunting for meat to sustain themselves and their families – normally they'd kill small to medium game like an impala, nyala, warthog or kudu. They would use hunting dogs to track the animals. Once the dogs had cornered the animal, the poachers would spear it and take the carcass out of the park. Sometimes, it was just young boys who seemed to get as much sport from running away from us as from hunting the animals.

How I cursed them as I dressed, before stepping out into the night.

Soon we were swallowed by the darkness, hardly lifted by the weakening sliver of moon, hanging so lonely behind the hills. A fiery-necked nightjar was calling, competing with a spotted eagle-owl.

As we moved in the direction from which we'd heard the barking dogs, I tried to clear my anger away. I knew my energy could be easily reflected back to us, that thoughts of violence can bring violence themselves. Instead, I tried to open myself to the intensity of the night, and open my senses to the dangers around us. I prayed for protection as we heard the snorting of a black rhino in the bush, the breaking of branches as it was feeding. Elephants were trumpeting nearby; we heard a leopard call, and a pride of lions calling and growling – I suspected that some were mating.

We crouched in thick bush, listening out for the barking of the dogs to guide us to the poachers. I looked up at the dipping moon saying goodbye, and the streaks of shooting stars, and felt my anger melt away. I listened to the calls of the spotted eagle-owl, now joined by a small scops owl, and my heart started singing.

I sat there, feeling the lingering irritation from my disturbed sleep washing away, as if the darkness was a river flowing through my soul. It brought an inner emptiness, a willingness to embrace openly what the night would bring. It was my job to catch these poachers, but now I was open to doing this without anger, with a gentleness of spirit.

After some time we heard dogs growling a few metres from where we were sitting. We were ready, and launched the chase. Their dogs barking at us accelerated the confusion, and in the chaos we managed to catch two poachers who'd been taken by surprise. Other poachers fled, but we didn't try to follow them.

The two poachers we'd caught sat under the tree, exhausted and covered in the blood of the animal they'd just killed. We sat with them, waiting for the sun to emerge. I realised that one poacher was the same age as me, and started a conversation with him. I asked him how many times he'd poached in the park – to my surprise, it was his first time. He stayed far from the park, but was visiting his uncle who stayed nearby. He was attracted by the beauty of this fenced land teeming with wildlife, and decided to follow his uncle to observe it more closely. He was not even aware of the danger he was putting himself in because he'd been so drawn by the beauty of the land.

My heart grew heavy as I listened to this young man. I thought of the many times when during foot patrols along the fence of the park, I had come across young adults peering through the fence, searching for the animals – not to kill, just to watch them. Their spirits seemed to hunger more than their stomachs. Most of the people whom I interviewed during this arrest, and others, wanted to work in the park. They felt that they were not being included or given the opportunity to help protect their heritage. To them, the park belonged to whites. Little wonder that they were poachers, for how can you protect something that you have never laid your eyes on? How can you embrace what has been stolen from you?

I remembered what I'd been told by the elders about the time when the game reserve was extended. There was tsetse fly in the region, and the government came to the people and said, 'You need to move because the tsetse fly will kill your cattle. Move over those hills there, where there is no tsetse fly.'

The people were reluctant, but their cattle were getting sick from the tsetse fly, so they agreed to move, and loaded their possessions on the trucks, sometimes paying for the moving themselves. When they got to the hills, they found that the tsetse fly was just as bad there. They tried to return to their ancestral lands, to the graves of their forefathers and mothers.

But the fences had gone up. The wild animals were on the inside, the people were on the outside. They'd lost their land and been given nothing in return. There were similar stories each time the park was enlarged. Like others in South Africa, the park had grown on a foundation of dispossession and dishonesty – and this history triggered anger in the locals. From the beginning, the story of the establishment of parks in Africa has been a story of exclusion.

The exclusion took many forms. Instead of being seen as partners to help with the conservation of our animals and land, black people were seen as being in the way, to be pushed aside and used only as labour, or perhaps as trackers. White scientists and conservationists often relied on the knowledge of local people, yet these people were seldom acknowledged or given due reward for their expertise. The name of Ian Player is well known, but less so the names of Magqubu Ntombela and the other indigenous conservationists who worked with him – although Player himself acknowledged the importance of their work.

For centuries black people lived with wild animals, without destroying the wilderness. Then white hunters came with their big guns and destroyed the animal populations within decades, hunting elephants and white rhino to extinction south of the Pongola River by the end of the nineteenth century. Yet this race, who'd killed all the big wildlife in their own country, thought they knew more about conservation than other races. Native wisdom was disregarded, creating anger and mistrust among the locals.

As the sun rose, we marched slowly towards the compound where a vehicle was waiting to take the poachers to the nearby police station. As I watched them being driven away, I thought of all the young men, like me, who had stood beyond the fence, looking with longing at the wilderness within. I thought again of that day with my father on the banks of the iMfolozi, which had sparked my desire not only to protect nature, but also to share my experiences of nature with my own people. My conversation with the poacher sparked a deeper understanding of where I had come from as a black child, and reignited my determination to find a way to open the path to the wilderness for my people.

———

A gunshot rang out, and a zebra fell – the animal was being sacrificed for the lion call-up that evening. This operation involved darting baited lions to collect data and blood samples, and checking their health status and condition. One of the lions had shown symptoms of illness, and we wanted to assess whether there was tuberculosis among the lions from eating infected buffaloes. I was excited: while I was familiar with lions in the bush, this would provide a unique, new opportunity to examine them up close.

I was lucky to be included. In the months of my volunteering, my passion and dedication had attracted the attention of other rangers and I had been drawn into many aspects of work, including rhino notching and buffalo branding and research. I had helped a team from Cornell University in the US with this research, and was shown how to enter data into a computer.

But lion call-ups were seen as 'white work'. Only the most privileged could participate. The white superiors would bring their relatives from the city, who would sit in the back of the vehicle, all wearing khakis and sunscreen, while watching the spectacle. So it was unusual for me to be involved, especially as a volunteer. But Dave Robertson and Craig Reed pushed to give me this opportunity.

Their superiors also agreed that I should be involved – perhaps because they recognised the need to transform the organisation, perhaps because a Portuguese film crew was coming to film the operation for a TV documentary and it would not have looked good for Ezemvelo if only white people were shown in the video.

The zebra carcass was laid open in the boma, with the contents of the rumen – the stomach juices – sprinkled around to create the scent of a kill. The scene was lit with infrared lights, and a recording of lions squabbling over a carcass was played over loud-speakers, drawing the invited guests (and uninvited visitors such as jackals and hyenas) to the scene.

I sat in the semi-darkness, holding my infrared light, waiting for the lions to come. It was an eerie scene: the red light, the sounds of bones crunching and lions snarling, the zebra carcass lying open and helpless. It made me realise how small we are as humans. To the lion, I was no different from the zebra. I was also food, and if I made any mistakes I could find myself on the menu.

The commotion and scent soon drew a pride of lions, approaching cautiously. One of the braver young males began eating. At the snap of the dart gun, the needle penetrated his skin. He growled and tried to bite the dart, then staggered. Drool fell from his mouth, and he collapsed helplessly to the ground. The recording was turned off, and we jumped off the back of our bakkie to chase away the other lions so that we could physically examine the darted lion and take blood samples.

As I touched the lion's fur, I remembered my first lion experience – how, when I had touched the paw print, that tingling had raced up my arm. Now I was actually holding the lion's paw, checking the teeth, holding the ears. As I laid my hand on his fur, I had the vivid sensation of two beasts meeting. My inner lion was meeting a wild lion, sparking feelings of intense jubilation. I knew the lion was sedated and that his feelings and consciousness were suppressed. But I was aware that we were connecting at a deeper level. It felt as if the lion could sense my energy as I communed with him and prayed for his survival from diseases, from poaching, from any of the stress and sadness that human society brings to wildlife.

I was so grateful to be able to embrace the lion as neither a hunter who wanted to claim his head or hide to hang on the wall of their lounge, nor a trader who wanted to export his bones and sell them for profit in Asian countries. I was meeting the lion as just another creature, jeopardising my own life for the survival of another species. I was a part of his species' life and together we embraced the deepest elements of survival, fighting together for one another's lives.

I sat holding the lion's paw as the roaring of the other lions and the whooping of hyenas under the moonless sky became a prayer. A prayer to commune with our wild animals, a prayer to protect our inner landscapes, a prayer to protect our inner wild animals. For the wild creatures within us are the custodians of our souls.

———•———

I had been at Masinda for some months, marked by the shift from winter into spring. New rains were bringing moisture to the air, and turning the decaying leaves on the ground to compost. Tender, pale-green shoots of grass were pushing through the red soil. The musical warbles of the Burchell's coucals could be heard every day, heralding the coming of a yet another rainstorm. The rains and spring winds were blowing away the dusty haze of the dry season, washing the sky a deep blue.

Spring's colours were slowly infusing the hills. The wild animals seemed to have a renewed beauty, a crispness to their markings, a new shine to their coats. There was a sense of busyness and celebration, as they feasted on the fresh, new shoots and prepared to give birth. Impalas and buffaloes capered about for the simple joy of it, and the air was alive with birdsong. My heart too was singing, with gratitude for the animals' beauty and their freedom.

A favourite spot to witness the rituals of the season was at Umthombo ka Ngomane, the spring of Ngomane. Ngomane was one of King Shaka's trusted advisers, who had been kind to him and his mother when they were in exile. Each year, as the rains soak the earth, the spring wells up, bringing life and hope to exhausted, hungry animals after a dry winter. The rains give birth to this and other springs – the rocks drip tears of fresh water, and the tears come together to form small, clear streams. Birds and animals congregate around these springs, grateful to quench their thirst with fresh, icy water after weeks of struggling at drying rivers and mud holes.

One afternoon, during our normal foot patrol, we waited near this spring, on the lookout for poachers. Spring was everywhere around us, and I sat marvelling at the revival of the world. The land had erupted in new flowers, pollinated by butterflies that had just unfurled crumpled wings as they emerged from their cocoons. Even the sky seemed to have been made anew.

A group of giraffes approached the stream tentatively, scrutinising the surroundings although there seemed no reason to be fearful. As they walked in slow motion their heads glided above the trees, like the masts of ships on calm seas. A flock of amahla-lanyathi (red-billed oxpeckers) clung to their necks, searching for ticks. The leading male paused for a few minutes, until satisfied there was no threat, then resumed the approach to the spring. I watched them carefully stepping on slippery, wet rocks, moved by the fragility of these huge creatures as they navigated the surface. I held my breath, to keep them safe from falling.

Next to approach the spring was a herd of buffaloes, with their woolly newborn calves. The sound of calves and mothers calling to one another disrupted the calm. Dust clouds rose from their hooves as they walked, a flock of cattle egrets flew above the herd, and all was bustle and noise. How different they were from the cautious, silent giraffes! The young ones ran playfully into the stream, squelching the mud, awakening my own memories of the simple, unselfconscious delight of being a child in a happy place. But while the young calves were drinking, oblivious to danger, their mothers were alert, scanning the bushes for predators.

The buffalo herd gradually drifted away, leaving their scent hanging in the air. Once again we could hear the sound of water dripping and trickling over the rocks, and the songs of the birds. Together they created a musical chorus that seemed to knit up the ragged threads of my soul into harmony. A giant kingfisher hovered, then dived – I could hear water splashing as it struggled to fly up with something in its beak. It landed on the tree next to us, and through my binoculars I could see it was a small frog.

A weeping boer bean grew nearby, its branches cloaked in bright-red flowers buzzing with bees. Turacos, starlings, sunbirds and bee-eaters flitted from branch to branch, the sunbirds' iridescent colours flashing in the sunlight as they took nectar from the flowers. A troop of baboons walked across the stream, and began turning stones over and stripping old bark from the trees in search of insects.

As I watched the animals coming from all directions to the spring, I felt a deep welling of spiritual harmony. It reminded me again of the abundant healing gifts of nature, asking nothing of you but your quiet presence, your attention, your patience and gratitude. As the land was awakening around me, I felt within me an invitation to live my life with an open heart, to live my life meaningfully in a place alive with meaning; to allow hope, love and joy to flood my soul.

We walked back to camp under the lonely setting sun, hovering above the horizon as if reluctant to leave. I felt so grateful for this renewal of life, for the dripping waters that brought new hope for the survival of this wilderness for another year. I felt again how the land was shaping my soul, the deep knowledge that without this connection I would feel lost, confused, alone and confined.

———————

Soon, spring gave way to the hot summer months, relieved by almost daily afternoon thunderstorms. Dark, heavy clouds were swirling in the sky as I followed Sonto up the hill on patrol one afternoon. We walked past a stand of bush medlars, my legs lamenting the steep slope, and sweat dripping from my temples. A hot wind from the east scorched our faces, and every millimetre of my body and my mind screamed in protest against the steepness of the hill.

Behind the rolling hills, lightning and thunder were embracing each other and giving birth to the rain. I paused to look back at the valley behind me. Impalas were pronking across the plains, the bright white patches under their tails flashing in the eerie light. They reminded me of ugcabhayiyane, the rain dances we used to do as kids, naked and unselfconscious, delighting in the cool showers.

A fresh breeze rose up, displacing the hot wind and soothing our hot, tired bodies. As we reached the top of the hill, thick drops of rain started falling.

The sweet smell of the dry ground embracing the rain floated through the bush. We ran under an umbrella thorn for cover, and stood beneath the spreading branches as they danced to the song of the wind. All around us, the African hills welcomed the rain, and the crickets broke out in applause.

The crack of thunder propelled us to move from the tree and the risk of a lightning strike. We followed a path winding through the umthombothi forest. A nyala barked at our approach; his second bark was drowned out by a blinding lightning flash and a deafening blast of thunder. The wind whipped up pitch-black clouds, darkening the landscape long before sunset. We were enveloped by two essential elements of life – fire and water. As we walked through the forest, I was pleased to see that the dry mud holes were filling with water – these holes not only provide animals with water but also serve as 'beauty spas' for the mud wallowers.

A row of elongated, slithering tracks told us of the struggle of an elephant to walk in the slippery mud. I tried to lengthen my stride into his footprints, but it was hopeless and I gave up, laughing. We could smell the elephant's fresh urine mixed with the water flowing into the nearby mud holes. He was not visible but his presence was strong, and that was enough for us.

The jagged forks of lightning prompted Sonto to divert us to a deep cave for cover. It was already occupied by a troop of baboons, which shouted at us as we approached the mouth of the cave – the smell of their faeces almost discouraged us from entering. Small babies clung under their mothers' stomachs and peeped out from under their arms. The alpha male took command of the troop and nudged it slowly to one side of the cave. We sat down quietly on the other side of the cave mouth. As the tension and the commotion from the baboons gradually subsided, the babies came out to peer at us from behind their mothers and fathers, or climbed onto their mothers' backs. The troop started grooming one another, and even began mating in front of us. We felt so honoured that they trusted us, as if we were two tribes sharing the space amicably, with no hostility. In sharing the space, we were remaking the connection between the human tribe and the baboon tribe, healing the wounds caused by human dominance and supremacy. The cave became a harmonised place of respect and unity, created through the principles of ubuntu.

A crack of thunder reverberated through the cave, accompanied by a blue flash of lightning. The alpha male called out in alarm, his shout echoing through the cave over the roar of the storm. The babies squealed with fear, as their mothers comforted them by allowing them to suckle. The wind blew heavy drops into the entrance, driving the troop further back. Hail started falling, lying in drifts at the cave mouth. The babies' curiosity overcame their fears, and they ran to the entrance, lifting handfuls of hail that they sniffed and nibbled. I laughed as they realised that it was tasteless, and threw it down on the ground.

A bright fork of lightning snaked through the clouds, followed almost immediately by another deafening crack of thunder. A pair of speckled pigeons fluttered around inside the cave in agitation. We huddled with the other creatures, united in our need to take shelter from the ferocity of the storm.

We had come to the cave uninvited, yet the baboons and pigeons had welcomed us. I felt again that deep sense of oneness that comes when wild animals tolerate you in their spaces, when you share spaces respectfully and without dominance. It was the same feeling I had when we drank in rivers where elephants were drinking upstream, or bathed in waters near where the buffaloes were wallowing, or foraged wild fruits where baboons and monkeys have foraged.

Sharing spaces is only possible if the animals are relaxed, if they accept and trust us, and it was this trust that made me feel so privileged. We too were trusting them not to harm us, and this mutual trust forged a deep connection. It revitalised ancient bonds with our wild animal brothers and sisters, bonds that have been eroded or destroyed by civilisation and by centuries of abuse of animals by humans. I sat in the cave with the baboons, awed by the power of the storm without, and by the quiet healing power of our interconnectedness, the sense of that sacred connection emerging from my unconscious.

To the outsider, I was a young man in an old cave with a family of wild baboons. But within myself, I was sinking into a powerful peaceful universe, where I was at one with everything, with the baboons and pigeons, with Sonto, with the fire and water of the storm, and all the rain-drenched hills.

———•———

One of our duties at Masinda was to monitor the transect lines. These are lines drawn across sections of the reserve that research-ers use to monitor samples of animal populations and activities. They are marked by natural features such as trees, but occasionally an elephant would knock over a tree, for example, and another marker would need to be found. Researchers from all over the world would come to study animal behaviour and populations in the reserve, using these lines.

One summer afternoon I was walking with Sonto along the Masinda hill towards the iMfolozi River. Clouds scudded across the sky, blessing us with brief patches of shade as they moved across the sun. As we passed a tall red ivorywood, we heard a dis-tinctive 'khle, khle, khle' coming from a small, brown bird with a tawny chest flitting about the branches.

'Do you know that bird?' Sonto asked.

'Yes, it's inhlava, the honeyguide.'

'You know it can lead us to honey?'

'Yes, my grandfather taught me.'

'Let's follow it …'

Sonto set off towards the bird. When it saw us approach it flew to another tree, and called again, 'khle, khle, khle', waited for us to approach, then flew to the next tree, and so it went on. We kept following it, praising it and calling, 'Ahah, ahah,' to encourage it. If the bird saw that we had lost sight of it, it would come back to a tree near us and call. I was concerned that we might get distracted from our tasks by this bird, for I knew from my childhood that they can lead you a long way. But just as I was about to say we should stop following it, it led us to a tall cabbage tree. Before we even heard the bees, we could smell the honey – as we drew closer, it was as strong as if we were sitting inside a honey jar.

We saw the dark hole in the trunk, and the bees flying in and out. The honeyguide was hopping from branch to branch of the tree, chittering in excitement. We were allowed to taste wild honey in the reserve, but not to burn leaves to smoke the bees, or break the hive, or do anything to harm the bees.

Sonto went up to the tree, and examined the hole. He said it was big enough for him to get his hand in without cutting or damaging the tree. He shone his torch into the hole, reached inside, and pulled out a creamy honeycomb, dripping with honey. The bees were buzzing around, but not attacking him; perhaps they knew he would take only a little. He reached back in and brought out another comb that had some bee larvae in the chambers – this he laid on a branch for the honeyguide, which flew down and began to feast.

We praised the bees and thanked them, then sat under the tree in the warm afternoon, enjoying our honey while our bird friend enjoyed the grubs. The gurgling of a nearby stream mingled with the chattering of the honeyguide, the buzzing of the bees, the wind whispering in the leaves. Across the stream, a herd of buffaloes lay in a field of gold and red grass, peacefully chewing the cud.

I thought back to the time when I had first met this clever little bird. My grandfather made his living by cutting and decorating fighting sticks, and selling them to new initiates and young men visiting from the cities. I had gone with him and another young boy to cut sticks. We'd grown bored of this and were playing in the water, when we heard the 'khle, khle, khle'.

'Do you know this bird?' my grandfather asked.

When we said no, he laughed and said, 'Iminqolo! You are mommy's boys, you don't know anything!'

He explained that the bird would lead us to honey, and started calling to the bird, encouraging it to lead us. The bird chattered back and, flitting from tree to tree, led us on a most horrific journey through the thickest and thorniest of thickets. At last we reached the hive, scratched from head to toe, although this hive had less honey than the one I found with Sonto.

By encouraging us to follow the honeyguide, wherever it might take us, I think my grandfather was teaching us to have faith in this bird. Such experiences were very important for helping Zulu children to build a connection to animals and to understand the significance of their clan's totem or spirit animal. If you have not run through the valleys barefoot following a honeyguide, it is harder to understand this connection. That is why, although there are many Zulu people carrying the names of their clan's totem animals, such as the Ndlovus and Nyathis and Ngwenas, that relationship is weakening. People are growing up in cities, away from nature, so those deep, protective connections are being lost.

Sonto and I licked the last drops of honey from our fingers, thanked the bees and the honeyguide again, and went back to our work of monitoring the transect lines. The bird chattered back, seeming to thank us in turn, and flew off into the bush.

———•———

Soon after this, I returned to Mduba to do relief work. One morning I was on patrol with Baba Thabethe, walking along a broad ancient wild animal trail. There was fresh white rhino spoor and fresh dung on the path, providing a feast for small flies and beetles. We spotted two red-billed oxpeckers chattering in the grass.

'Qaphela phansi amahlalanyathi' (Be careful, the red-billed oxpeckers are on the ground), Baba Thabethe, our patrol leader, warned us. This saying means that the oxpeckers may be on a big animal that is lying down, hidden from our view by the grass, as oxpeckers are rarely found on the ground.

Oxpeckers usually ride on big mammals such as buffaloes, rhinos, elephants and zebras. They warn the animals of our presence, but they also carry warning messages for us rangers. His words made me think of the honeyguide, and I realised that here was another experience of a mutual understanding between us and the wild birds.

We could hear the twittering of the oxpeckers growing louder, but could see nothing through the tall red grass until a breeze moved the grass, revealing the top of a heavy head, two ears twitching … The birds were indeed resting on an animal: a white rhino.

He caught our scent and scrambled to his feet, lifting his razor-sharp horns as he sniffed the air. He ran around, sniffing the ground and twitching his ears, followed by the faithful oxpeckers. Flies buzzed around his small eyes, set deep in wrinkled folds. Thorns scratched his thick skin as he walked through the sickle bush, still sniffing the ground. We followed him slowly, along an ancient trail forged by many creatures before him. He spent his time sniffing, scratching, peeing and defecating. As soon as his dung landed, it was flooded with flies and dung beetles. Some birds flew down and scratched the ground nearby, hunting the insects that had come for the feast, or searching for seeds or other food in the dung.

Each time the white rhino turned his dung with his horns, more life revealed itself. The smell of fresh dung attracted more flies and dung beetles; millipedes and centipedes crawled out from underneath the dung, to the delight of the birds feeding on them. A great white egret followed, snapping up the insects that the rhino disturbed as he brushed the bushes bordering the path. I was fascinated to see all the ways these animals connect and give life to one another. While there is competition between animals for food, so many of their interactions that I had observed were more about cooperation and sharing. Baba Thabethe and Dumisane Khumalo followed this principle, sharing food willingly with me, and always reminding us to leave enough for the baboons and other animals if we picked wild fruit from the trees.

The birds on the rhino's back chattered more loudly as he came near us, warning him of our presence and warning us of his approach. We slowly retreated, walking backwards and trying to be silent, but the sound of the branches scratching our clothes alerted him and he ran towards us. We knew he wasn't charging, just trying to escape any danger – but his hump was up and his sharp horns were aimed straight at us. We took cover behind the trees as he pounded towards us, his footsteps shaking the ground like a small earthquake. I was horribly aware of my own vulnerability, recalling my terrifying encounter with the black rhino.

The white rhino stopped barely a metre away, his small eyes fixed on my tree. I froze, hardly daring to breathe. I could see his eyes blinking and the ticks on his ears. He sniffed the air, and something in him changed. He ran aimlessly around and lay down on the ground right in front of me. Birds landed on his back, eating ticks off his skin and removing crusts of mucus from his nostrils. He kept blinking as flies crowded around his eyes; his stumpy tail flicked from side to side. I watched this great, powerful beast lying calmly right before me, moved (and relieved) by his recognition of my peaceful intentions.

This made me realise how much of our communication with animals happens on a deep and subtle level; that with their heightened senses animals can look into our souls, sense our emotions and intentions, and reflect them back to us. If we are peaceful, they will reflect that peacefulness. I had noticed that when we went out in a vehicle to cull impalas, they were nowhere to be found, and yet if we walked or drove out with no intention of killing them, they were all around us. They could even read our intentions when we were in a vehicle. I was learning that nature is a mirror I can use to reflect on my own intentions, for introspection into my emotional state, revealing feelings that may have been hidden from my conscious mind.

Watching the white rhino bull's comical behaviour brought out a playfulness in me that helped me step away from my anxieties. Connecting with animals brings us out of ourselves, helps to release us from the prison of the ego, from our need to be successful and important. Instead of focusing on ourselves, we can celebrate our small humble part in this great chorus of being, and let the songs of life and freedom reverberate through our hearts.

───────

The sound of someone humming and whistling brought me into wakefulness. My rusty old friend (I was staying back in my old room at Mduba) complained as I wriggled into a different position. Through my window I could see the masked weavers busy at their nests in the umbrella thorn. The small yellow flowers had started to bloom, and pale fragile leaves were erupting all over the tree like unshaven stubble.

My colleagues' banging doors warned me that it was nearly time for patrol. I jumped out of bed and opened the door, relishing the fresh new air, the twittering of the weavers, the smell of cooking phuthu.

I set off some time later with Dumisane and Baba Thabethe, on a gentle path that wound through the acacia forest before zigzagging along the stream. A lonely male kudu paused in his browsing to contemplate us briefly, a fresh sprig of leaves hanging from his mouth. I watched as he moved off gracefully through the bushes, bearing his magnificent spiral horns. The fragrant scent of the woolly caper bush hung in the air. Deep inside its branches, a scarlet-chested sunbird was singing, joined by a cloud cisticola nearby.

The sun rose higher, beating on our backs as the path took us up over the hills through fields of red grass with scattered bushwillows stretching to the horizon. The archaeological remains of an ancient blacksmithing site indicated the age of the path we were walking on. The path has also been used by animals for centuries, the rocks worn smooth by the passage of hundreds of hooves, pads and paws over the years. As we walked on the ancient path that had been forged by generations of wild animals, I felt it connecting me to my human ancestors. Had they walked this same path, when they lived off the land hundreds of years ago?

We passed a stump used as a rubbing post, its surface polished to a high sheen by generations of animals rubbing against it. There was a dead tick lying by the post, its wrinkled body like a deflated balloon. I noticed the tracks of many birds, drawn by the ticks that were rubbed off on the post.

We reached the top of the hill and looked down at the White iMfolozi River, its silvery water snaking on golden sand. A lone buffalo bull waded across, pulling a V of silver ripples along the water's surface. The wailing of trumpeter hornbills echoed across the forest.

We wound down towards the riverbank. My muscles were screaming for a break after the long climb, so I was relieved when Baba Thabethe called a halt. We stopped under a huge umkhiwane tree, its yellow bark dripping milky, sticky sap from a scar scraped by an elephant tusk. The trunk was so massive it would have taken five men to circle it.

I sat resting my back against the tree. A deep, sacred stillness lay over the land, broken only by the soft sound of small unripe fruits falling and the buzzing of insects feeding on them. My spirit floated through the branches, as I recalled the stories told by my father of the mighty Imfolozi rivers before Cyclone Domoina in 1984. He spoke so often of the beauty of the umkhiwane forest along the riverbanks, before the cyclone uprooted them – but this ancient tree must have been one of the few survivors. I imagined its giant roots probing the earth beneath me, as I enjoyed the coolness of the shade and reflected on the generosity of the tree in giving life and shelter to others. How many creatures and humans had it helped to sustain? Without trees, so much life would cease.

As we were taking a long break, I decided to take off my clothes to bathe in the river. As I strolled naked over the warm, gold sand towards the water, fresh leopard spoor gave me a shiver – I glanced around, humbled by the reminder that I was just another meal to many animals here.

I waded into the river under the warm Zululand sunshine, feeling the silky water flowing over my naked skin. We wear clothes without thinking, yet they are a constant barrier, an artificial skin that disconnects us from the world beyond. Now, without clothes, there was no barrier between my skin and the world outside, nothing to stop me from entering the divine state of oneness with the earth. I felt the simple completeness of my animal self. Without clothes, we are no more or less than any other creature; nakedness is a refuge of equality for all living organisms on earth. We come into this world naked, and to revisit that space of nakedness in the wilderness becomes an orgasm of the soul.

I leaned back in the water, feeling the soft mud between my toes, and the cool wind blowing through my hair. I felt the worries of my life slowly detach themselves and float away on the current. I thought back to my very first patrol nearly three years before, when I had walked naked on the banks of this river. Much was still unresolved ... I still had no clear prospect of a job. My family was struggling, and we desperately needed funds. But I had been held through these difficult years by so many good, kind people – by Sabelo Msweli, Dumisane Khumalo, Nkalakatha Nxumalo, Baba Thabethe, Craig Reed, and Dave and Erica Robertson. They had shared their deep wisdom and knowledge about the wilderness, shared their food and resources, given me opportunities to learn and experience so much, taken me seriously, encouraged me when I'd felt disheartened.

With their support, I had been able to step into the wilderness. I had indeed been able to weave myself into the rich tapestry of life that flourished in the iMfolozi, to feel that sacred sense of oneness with the earth and all the creatures that crept, galloped, scuttled or slithered on its surface, from the roaring lions to the timid chameleons – oneness with the birds and insects that filled the sky, oneness with all the plant beings that brought life to the animals, from the towering sycamore figs to the delicate grasses.

I was floating on gratitude, for this kindness, this richness. I lifted my face to the brightness of the day, letting it soak into me and brighten the luminosity of my inner life, which I knew would bring me light during life's darkest moments, whatever lay ahead.

———•———

Not long after this, a chance meeting led to a great change in my fortunes. As we were coming back from patrol on a very hot day in October, my section ranger radioed us to help someone whose minibus had run out of fuel. A tall man with long hair emerged from the bus. He was deeply tanned from hours in the Zululand sun and, to my young eyes, had the rugged and exotic look of a rock star. But the logo on his sunhat told me that he worked for the Wilderness Leadership School, and the scratches from thorn trees on his arms told me that he spent his days in the bush and not on a concert stage.

As I was pumping fuel into his petrol can, he remarked, 'You look young to be working here.'

I explained that I was volunteering, not working. He looked at me appraisingly and said, 'Well, now you have a job.' Then he filled his tank with the petrol and drove away.

I wasn't sure what to make of this, but a week later I was called to send my CV to the Wilderness Leadership School, followed by an invitation for an interview. Whatever I said must have convinced them, because soon after that they told me I could start my training for NQF Level 2 in Guiding – a National Qualifications Framework certificate.

Before the chance meeting with the man from the leadership school – whose name I would later learn was Paul Cryer – my ambitions had been to go further with law enforcement, to become a section ranger. But law enforcement was never my passion. When Paul brought this opportunity, I realised that wilderness guiding could be as important to conservation as being a section ranger – and it fitted my life's vision better. Again, it seemed that the universe was showing me which path to take.

And so, one warm day the following April, I presented myself to start my training as a wilderness guide at the Wilderness Leadership School headquarters in Durban. The tough days of working without pay were over.

Part Four

—•—

A NEW DAWN IN THE WILDERNESS

2002 to 2007

THE DENSE FOLIAGE OF THE UMTHOMBOTHI TREE ENVELOPED OUR group, its branches spreading wide to welcome us as we declared it home for the night. I dropped my backpack, and stretched, relieved to be free of the weight, my heart singing with gratitude. After three tough years of volunteering, I had awoken to a fresh new dawn in my life. As a trainee guide, I was free of the daily routines of the foot patrol and had been given the opportunity to explore nature and myself, to learn from the best in the field. Best of all, I was to receive a decent stipend during the training – at last, I could be free of the constant worry of where my next meal would come from, and could even help to support my family.

This was my first night of being on a wilderness trail, and I was eager to drink in every experience. I looked out across the floodplain to a breeding herd of impalas grazing and pronking, some snorting and cautiously watching us. I wandered down to the river, taking in the two buffaloes lying peacefully on the sand, the graceful curves of the Goliath heron fishing in shallow pools, the white rhinos coming from all directions to drink, their calves stumbling and mock-charging, ears twitching. Giraffes drifted across distant hills, nibbling the soft new leaves that had sprouted on the black monkey thorn trees. Falling leaves sang to the southerly wind that blew across the valleys and hills, harmonising with the calls of the trumpeter hornbills. This was paradise, and I was in it.

I picked a handful of wild basil leaves and crushed them between my fingers to release their sweet scent as I walked slowly back to the group, now gathered under the umthombothi tree. The giant web of a golden orb spider stretched between the branches, its strands gleaming gold in the sun. Its victims were carefully hung, wrapped like well-preserved mummies; the huge female was suspended in the centre of the web. On the far end a tiny male crouched, perhaps hiding from the female in fear of the cannibalism that usually follows mating.

The last of the sunlight caught the sand banks of the iMfolozi

River, and turned the water silver. A giant kingfisher hovered in the air searching for a meal, its wings spread over the water. I prayed for it to have luck – I knew all too well what it was like to go to bed hungry. Again and again it dived and came up with nothing, but at last I saw a flash of silver in its beak, and it flew away with its meal.

When darkness approached, my trail leader instructed me to kindle the fire. Soon the warm smell of dry sage burning permeated the air. I stood watching the pale wreaths of smoke whirling into the sky, and thanked the spirits of the place for allowing us to be present, to warm ourselves by the fire. The group gathered around the blaze, drawn by the flames and the aromatic scent of burning wild camphor bush. We sat quietly, watching the kettle heat up, absorbing the simple powerful medicine of a campfire in the

At the campfire – an invitation to be present

wilderness. This ancient ritual reaffirms our ancestral connections, opening our inner senses to the clear and present messages from nature, opening doorways in our hearts and minds to the wisdom that comes with deeper searching, deeper knowing and deeper understanding. Simplicity is the alchemic substance that brings transformation in nature – simplicity and showing up, being fully present. The campfire is an invitation to be present in this simplicity.

Soon the sounds of the night creatures began, background music as we prepared supper – the barking of the bushbuck in the depths of the umthombothi forest, the mournful calls of a bush-baby in a nearby sycamore fig. As I gazed into the dancing flames, I felt a deep and ancient sense of belonging. I knew I was where I should be, and that my life was taking me onto the right path.

———

The Wilderness Leadership School was founded in 1957 by Ian Player and Magqubu Ntombela. Their vision was to give people an immersive experience of walking in the wilderness and sleeping under the stars for several days, and to use this experience to teach them about conservation, teach leadership skills and ignite a passion for nature. Youth were especially targeted. The early trails run by Ian and Magqubu were quite different from the wilderness trails of today – they used donkeys to carry equipment, and would pitch tents and shoot animals to provide supper, but these practices were later dropped. I can just imagine how today's trailists – many of whom are vegetarian – would cope with shooting and skinning an animal!

I spent three years in training, and another two as an intern, conducting trails under supervision. My mind and senses were wide open, and I was ready for anything the experience might bring.

My home base was at the Wilderness Leadership School campus in Durban. Although located in a city, this great sprawling compound was surrounded by forest, where zebras, impalas and bushbuck roamed freely. I was pleased to be sharing a room with Mandla Gumede, who'd been with me at the Masinda camp while I was volunteering – Mandla was a good friend, and helped me to face the new challenges. He was my personal comedian, softly spoken but always quick with jokes that would keep me laughing for hours. We pulled each other through our learning: when he got his driver's licence before me, he encouraged me, sometimes letting me drive when we were in a car together. When I got my first certificate in computers, I helped him develop that skill. I was grateful to have such a good friend taking this journey with me.

We had some theory classes in Durban but, in line with the philosophy of the school, the emphasis was on learning through observation, and the best opportunities for observation were on trail. Whenever there was a space, the trainees would go along and learn as one of the trailists. So I continued to spend much of my time on trail in the iMfolozi, and this wilderness continued to be my most powerful teacher.

However, being in Durban did give me the opportunity to understand the logistics of getting tourists from the airport in Durban to the iMfolozi, catering and buying food, organising and cleaning equipment. This was valuable experience in teaching me about the administrative side of trail management. In my spare time, I could explore the city. Paul Cryer often came with us, and helped to open my eyes to some of the ecological impacts of the city. I remember once when I commented on the refreshing cool air in the shopping malls, he explained how keeping these malls air-conditioned was accelerating climate change. Every moment with him was a learning opportunity, but he also soon became a good friend.

We were privileged to learn bush interpretation from extraordinary and highly skilled trail leaders like Paul, which enabled me to build on the knowledge I'd already acquired from my years as a volunteer. We learnt to distinguish a spider track from that of a millipede, to tell a tree browsed by a black rhino from one browsed by a kudu, to read an animal's mood from its tracks. Our ears grew sensitive to the myriad different bird calls, until we could identify birds by sound or by sight. In the evenings, I would ask the trail leaders to go into more detail about what we'd discovered that day, so that I could record this information in my journal.

Watching these skilled leaders interpreting bush signs was like watching a dance, an artist at work, and the learning was enriched by the wonder and respect they felt for the creatures they were observing. Unlike the seminars at the school campus, bush training had no schedules or routines, which encouraged me to learn more from the heart. I far preferred learning this way, instead of just reading bland, dry textbooks!

In the evenings we held wide-ranging discussions around the fire, covering socioeconomic issues, racial issues, global warming, economic problems, the global environmental crisis, spirituality … Back in Durban, we were given books to expand our understanding of these issues. A whole new universe opened up for me, helping me see the bigger picture and understand that the work we were doing was part of a global effort to value, promote and protect nature. I developed tools for reflecting on the world and on my place in it. This stretched my understanding and knowledge far beyond what was usually required of a trail guide. On the days following these discussions, I was grateful for the long hours of silent walking, which gave me time for introspection.

We also learnt many practical skills, such as finding our way around computers, driving, and handling a rifle. I embraced the discipline needed to acquire these as I knew that with enough effort these skills would become part of me, and would always be there to help me through life.

Acquiring such skills brought responsibility – and we soon learnt that the wisdom of knowing how and when to use a skill was as critical as the skill itself. This was especially true of rifle handling – for what greater responsibility is there than holding the power to inflict a mortal wound with your finger? Sabelo Msweli had taught me *how* to handle a rifle when I was volunteering. But Paul Cryer and Keith Robert, a seasoned conservationist and firearm expert, deepened my understanding of *when* to use it, because they were so passionately committed to preserving the lives of animals. They hammered into us that the rifle should only ever be used as a last resort, never as a first option when confronted by a dangerous animal. They told us to imagine wild animals as our brothers, to walk into the wilderness armed with peace and respect, rather than relying on a firearm. They said that an attitude of peace can be as powerful as any weapon, for wild animals can sense your intentions. If you approach them with peace, you can avoid bloodshed. Wild animals can feel your emotions, just as the dogs or horses at your home understand your thoughts and feelings.

These insights resonated with my own experiences as a volunteer, and brought to my mind the words of my grandfather at my ukuthomba ceremony – that good warriors walk with peace and harmony as their spears. My own experience as a guide has confirmed this lesson again and again. When we are at peace with our surroundings, all melts into oneness, and peace flows like a river through all forms of life around us. Peace is more powerful than any weapon on earth. If only humankind could learn to use this power!

———•———

On one of my earliest trails with the Wilderness Leadership School, I was to learn again of the deep layers of connection that may be forged between humans and animals.

It was a dry winter's day as we made our way along the Mhluzi Stream towards the Black iMfolozi. (The name Mhluzi means 'soup'. It was named after a vicious battle between Shaka, with his Zulu troops, and Zwide, leading the Ndwandwe tribe, caused the stream to run red with blood, and it was said that the crocodiles were drinking blood soup.) The mellow call of the brown scrub robin in the reeds took me back to my childhood days in the bush. My old rucksack dragged on my shoulders; the sweat on my legs irritated the itching and the sting of old tick bites.

We saw the buffaloes as we approached the Black iMfolozi. They were walking in the middle of the river, or resting on the sandy banks. Alerted to our presence, they began to move restlessly. The calves scrambled about in confusion as the herd began to move away from us.

I was walking up front with the lead guide when a woman called me from the back of the group, asking for assistance – she was struggling with the demands of the walk. I called back to say that I was coming. At the sound of my voice, the buffaloes paused and lifted their heads, sniffing the air. They stopped their retreat, and started to drift closer to us.

I began to tell the group about my experience of looking after buffaloes as a volunteer. The sound of my voice again seemed to draw the animals, until they were all around us, the bellowing of the calves and mothers resounding in my stomach. The trailists began to get uneasy. But we were on the edge of a donga looking down at the animals in the river, so it was not easy for them to reach us. And they were not behaving aggressively.

'Do you think they've smelled a predator?' our trail leader asked. 'They're acting so strangely.'

I looked down at the restless mass, now just below us. It was then that I noticed the brand mark on their flanks, and burst out laughing.

'But these are *my* buffaloes!' I said.

As a volunteer, I had spent three months working on a TB research project with buffaloes at Ezemvelo. TB is of great concern

– infected animals are more likely to be eaten by lions, as the illness weakens them. TB then gets passed on to the lions, and can decimate the lion population.

As part of the project, the buffaloes were rounded up and herded into an encampment of different bomas. They were darted and, while sedated, had blood samples taken, their teeth inspected and measured, and their age assessed. Then they were branded to show that they'd been checked. I helped with taking the samples and the branding, and I looked after them while they were kept in the boma – some animals stayed for two or three weeks.

I talked to them as I went about my duties of refilling hay nets and water buckets. I had known them to be aggressive creatures – in conservation circles they've been called 'black death' because of their tempers when they are disturbed, and they compete with the black rhino for being the most feared and dangerous animals to encounter in the bush. I kept a safe distance from them, but over time I could sense a connection developing. Whenever they saw or heard me approaching, they would move towards the fence and follow me. This was very touching, especially considering their reputation for aggression. At night, lions would prowl around and scratch the metal walls of the boma. I camped in a tent beside the buffalo encampment, and would call out to chase the lions away. I think the buffaloes began to recognise me as their protector.

I began to develop a deep respect and love for these animals. They are powerful social animals, incredibly brave, and will go to any length to protect weaker members of the herd. When threatened by lions, they form a circular wall around the young and the sick. If any buffalo calls for help, others will come to its aid, risking their own lives. As I moved among them, I could smell their sweet, grassy breath and feel their collective heartbeat, as if their hearts were beating in my chest.

The hardest part of this project was when the infected buffalo were culled. Seeing them being shot killed something in my soul, especially as they had seen me as their protector. I tried to look at it as a conservationist. I did not support the culling but I did understand the need to revitalise the health of the lions. Balance is vital in nature – and the health of top predators is critical to keeping that balance. The truth is that even in a big wilderness area like the iMfolozi, human fences mean that populations are artificial. Animals cannot roam freely, so we have to make these kinds of interventions to keep as much of a natural balance as possible. I knew that killing the infected buffalo would save the lions, but each time one was killed, it ripped through my heart.

But my heart sang now when I realised that these buffaloes still recognised my voice, all these months later. All the discomfort I'd been feeling on the walk vanished. Here were these dangerous beasts, recognising me as their brother and friend. The trailists too were deeply moved, as I discovered when we reflected later.

'I was afraid,' one said, 'but my fear helped me understand something about myself, helped me find the courage to be more true to myself.'

It was only several years later, however, that I discovered that I had a deep ancestral connection to buffaloes that I'd never known about. Many years after this happened, I spoke to a researcher who specialises in the origins of clan names. He told me about one of my early ancestors – Sontshikazi, son of Ndaba. He was a powerful, wealthy man with many cattle, and many wives, and was also famous for his work as a sangoma or diviner. Sontshikazi always wore a full buffalo hide over his shoulders, like a mantle. His family followed this practice, wearing a full cowhide if they didn't have a buffalo hide. When the neighbouring clan noticed this dress code, they started calling Sontshikazi's people amaMbatha (those who clothe themselves with hides). The name stuck, and Sontshikazi's homestead became known as Emambatheni (the place of the amaMbatha) and Sontshikazi's children were called the children of Emambatheni. And so the Mbatha clan was born.

The buffalo, my clan's totem animal

This researcher explained to me that the Mbatha do indeed have a totem animal, and that this animal is the buffalo, although they are also closely linked to the mountain, which I knew already.

Perhaps there was also some sense of this deep ancestral connection between me and the buffaloes. As they were my clan's totem animal, I had an ancestral obligation to protect them, and perhaps at some mystical level beyond our understanding the buffaloes and I had recognised one another.

———•———

A few months into my Wilderness Leadership School training, I was consumed by horror on hearing that one of my former Ezemvelo colleagues had been trampled to death by an elephant while on trail. He was a guide in training, working with Ezemvelo KZN Wildlife. We had worked closely together, and I was deeply shaken by his death. We spoke at length with our mentors, trying to process what had happened to this young guide – they helped me to understand it better, and helped to calm my fears. Like all trail leaders, the guide had had a loaded rifle ready in his hands but no time to use it. This showed us with painful clarity that a gun is an imperfect weapon for managing dangerous animals. But it was only some years later, when I learnt of all the circumstances, that I realised that this tragic incident also held a critical lesson about how to keep yourself and your trailists safe in the bush.

I was on trail in Jetzendorf in 2017, while travelling in Germany, when I met a German who'd happened to be on that same trail when the guide had been killed. He told me that they had been watching four elephants, and did not notice a fifth one coming through the bush – it was this one that had killed the guide. What this incident starkly revealed is that a gun on its own will never keep you safe, and can even endanger you. Because the most dangerous animal is not the one facing your gun. It is the one you haven't seen.

I had learnt this lesson as a very young boy when I'd lost my best friend to a crocodile; we were to learn it again and again, throughout our training, and one of our greatest teachers was the Black iMfolozi River.

This river might have looked calm and tranquil, but it was not for nothing that the elders called it Imfolozi Emnyama inkethabaweli (the river that chooses who crosses it). The reeds on both banks gave perfect cover to a lion or a buffalo. The deeper channels most often harboured crocodiles, but it was difficult to gauge the depth of the river as its opaque waters were impossible to see through. We quickly learnt that these treacherous channels often flowed on the sides of the river.

While I was a still volunteer, we witnessed a male lion being taken by a crocodile. One warm afternoon, as we approached the river, we heard the male calling. I praised him as my father used to do, calling him iNkosi Yehlathi, the king of the bush. As we came onto the sandbank, we saw him on the far side of the river, walking slowly while sniffing the ground, pausing now and then to stare into the water – perhaps seeking a shallow channel where he could safely cross. At last he began to wade across, but he was not happy. His tail was up, and he was moving fast, looking around on all sides. Suddenly, there was a tremendous splash as the lion was pulled down. The lion's snarling face appeared and disappeared below the water, the tail of the crocodile swished and slapped the water, and fountains of blood shot up from the water. After some minutes of deathly struggle, the lion succumbed. As the crocodile spun the carcass to rip it up, the lion's tail spun and smacked the water as if someone was hitting the water with a stick.

Less than one hundred metres from where the lion was taken, we witnessed another crocodile attack on a lion three years later, while I was in training at the Wilderness Leadership School. It was about midday, and we were sitting on rocks overlooking the river. Three males and one lioness, followed by four cubs, were crossing the river, about eighty metres upstream from us. The female started leaping around, and slashing at the water as if she'd spotted a crocodile. All went quiet for a few seconds. Then the last of the four cubs was pulled under the water with a loud splash. The other three scrambled frantically after their mother, but their brother was lost.

So it is in nature – we walk, always, the line between life and death. Knowing this can keep you awake, keep you sharp and vigilant, and literally keep you alive. If you are not fully alive to the world around you, you could risk losing everything: not only the profound sense of wildness, the experience of harmony and oneness with your surroundings, but also your life.

Connecting fully with your bodily sensations, and with all the subtle messages of the wilderness – the trembling of a leaf, a certain stillness, the tone of a bird's call ... all of this creates a oneness with your surroundings that is deeply healing. But it is also your best defence in the wilderness, and will save your life long before the gun in your hands does – without spilling a drop of blood.

During my Wilderness Leadership School training, I had the privilege of working with many powerful mentors. But one who left a very deep impression on me was Baba Gumede.

While I was a volunteer for Ezemvelo, Baba Gumede had conducted trails for the Wilderness Leadership School. When I was interviewed for a position at the school, I was told I would be stepping into Baba Gumede's shoes, as he was getting older and finding the demands of constant trails too tiring. These were very big shoes to fill.

Baba Gumede had been working for the school since its beginnings, and had worked closely with Ian Player and Magqubu Ntombela. He was a truly humble man, with a vast and deep knowledge of the wilderness. He had been held back only by his limited confidence in speaking English, but he would often lead trails with a backup guide who could translate for him.

Although I'd been told that Baba Gumede was wanting to retire, when I met him I was surprised by his strength and energy. From the beginning, he took me seriously and gave me so much respect. In turn, I could recognise that here was a man who had much to teach me. So I was very pleased when Paul Cryer assigned Baba Gumede as a backup guide to help me and Mandla Gumede (no relation) when we started conducting trails as interns.

The respect that Baba Gumede gave us, his humility and his profound respect for nature all made me instantly fall in love with this man. Every day, I was astounded by the depth of his wisdom – in particular, his vast wealth of indigenous knowledge about the plants, animals, weather, geology, astronomy and history of the region. He helped me when I wrote my exams on wildlife management, as there was a component on indigenous systems.

When we were on trails together, I gave him many opportunities to lead, acting merely as his interpreter, which he deeply appreciated in turn. Indigenous knowledge is not highly prized in conservation circles, and we were expected to regurgitate our textbook knowledge of the Latin and English names of the various species, rather than the isiZulu names. Some of the trainees who had been to university tended to look down on those of us who hadn't, believing that their book knowledge was superior to ours. This attitude, combined with Baba Gumede's limited English, meant that his expertise and deep knowledge were not always recognised and valued. But I knew that no book learning could replace a lifetime spent in the bush, learning with your own senses and your heart, and Baba Gumede's wisdom far surpassed that of many of the most educated scientists. Just as I appreciated the respect he offered me, so he was appreciative of the respect I gave to him, and the space I created for him to share his insights.

It was not only wisdom that Baba Gumede offered. Whenever we came across a dangerous animal such as a buffalo, the normal practice was for the lead guide to wait in the open while the backup guide got the trailists behind a tree. Baba Gumede would get the trailists behind the tree but would not hide himself – instead, he would stay out with me. When I asked him one day why he did not hide behind a tree, he said, 'In my blood, it doesn't feel good to leave you with the animal ... What will I say to your father if the animal kills you? I am here to protect you as my son, to see that you do not get hurt by these animals, until I see that you are strong enough to face them. This is only our third month working together, so I still need to protect you.'

That night, I gazed at the stars and thought about how lucky I was to have someone care so much about me.

He felt the same respect for animals as he did for humans, and always said, 'If you can point a firearm at an animal, it means it could be easy for you to kill a person too.'

I remember one night when I was sleeping and an elephant approached the camp. In my sleep, I heard him talking to the animal,

Baba Gumede, holding a photo of himself with Ian Player

but I thought I was dreaming. He was saying, 'Umfowethu [my brother], we are here for tonight. This is your place, we know, but please can we share this place with you? We are here with deep respect. We are not here to make you angry but we are here to embrace your place. You are bigger than us and we respect you.'

After some minutes, as if he understood Baba Gumede's words, the elephant turned and walked silently away.

The next morning, the people asked me what he was saying to the elephant. I asked him, and he explained. I said, 'I heard you talking, but I thought I was dreaming.'

He replied, 'Yes, you were asleep and I did not want to disturb you because I wanted you to catch a good rest for the next day.'

This was the man he was, and he has remained a trusted friend and mentor throughout my life.

———————

About eighteen months into my training, we were taken for navigation training to the nearby uKhahlamba-Drakensberg, South

Africa's highest and longest mountain range. Like the other guides from the area, I had no need to navigate in the iMfolozi. My years as a volunteer there, and the weeks spent there with my father as a child, made the area as familiar to me as the contours of my own face. I knew it by sight, smell, touch and sound. I had traversed it at night, in the mist and through driving rain. I knew where to find shelter, where the rivers were safest to cross, which thickets and reed beds were most likely to conceal dangerous animals, where animals would come to drink when the rivers and waterholes ran dry.

But the Wilderness Leadership School was interested in exploring wilderness trails in other parts of the country, and wanted us to learn how to navigate unfamiliar territory using maps, compasses and calculations in place of our own knowledge and intuition.

Because we'd always been in districts that we'd crossed and recrossed a thousand times on foot, using a map even for simple navigation was new for us. And this was not simple navigation – we had to read contours, calculate coordinates, use a compass and a map to find truth north, and apply these skills to locate points and topographical features marked on the map. We had to look at the waving contour lines, scribbles and arcane symbols on maps and conjure them into the hillsides, crevices, boulders, forests and rivers around us.

Acquiring these skills was very tough, requiring a level of mathematical skill that went far beyond what we had learnt at school. Many of my colleagues were defeated; some did not even attempt it. But for me, the stakes were high.

It was the early 2000s and, like other institutions, the Wilderness Leadership School was keen to transform the organisation by giving opportunities requiring more skill to black employees. I had been identified as one of the first black guides with the potential to lead 'meaningful trails' – trails conducted to help with healing historical divides such as those in South Africa or Northern Ireland, trails aimed at truth and reconciliation, trails where the participants were seeking more than just an exciting adventure in the bush. The Wilderness Foundation from Port Elizabeth (now Gqeberha) was also seeking to transform the organisation, and had approached me to lead trails for them as well, under the umbrella of the Wilderness Leadership School. Many people were relying on me to develop the skills I needed for this work, and I was under heavy pressure from Paul Cryer.

Paul was tough on me. He was far more lenient with the other young trainees, accepting excuses from them that he would never accept from me. At first, I began to think he disliked me. But then I noticed his delight when I would do something well, and I began to think that maybe he was putting pressure on me to bring out the best of my abilities, like you squeeze a walnut hard to get to the best part of the nut inside. He expected more of me because he believed I was capable of delivering more. When I realised this, I felt extra pressure to meet his expectations.

As I struggled with the complexities of navigation theory, I was so afraid of letting these people down, and of letting myself down – for I knew that many paths would open for me if I could succeed. My fear of failure banished sleep and tormented my mind so that it became even harder to concentrate on the tasks.

The turning point came one evening, when we had to do night navigation. Most of our tasks involved locating obscure features such as unknown caves during the day. But this was a specific night navigation task: somewhere on the dark slopes of the uKhahlamba mountains – the isiZulu name for the Drakensberg – a slab of chocolate was hidden, and it was our duty to find it. The object we were seeking might make it seem like a flippant exercise, but for me it was deadly serious. If I could not bring that chocolate bar back, I would fail the course. I would not get the certificate, and would let my leaders and myself down. It would paralyse my dignity, paralyse my chances of qualifying as a guide, and all my opportunities would slip through my fingers.

The mist had already descended when we set off in the darkness, me and a colleague from Kenya, armed with a map and a compass. Through the misty darkness I could hear the yowling of black-backed jackals across the valley, the rush of nearby waterfalls gushing over soft dolerite rocks, the eerie whistling alarm calls of the mountain reedbuck. I had a Betalight torch – these were high-quality, reliable flashlights, famously used by astronauts, but mine was old and cracked. The map was in a plastic Ziploc bag, which was fortunate, because a heavy storm soon followed the mist and in no time we were drenched by a downpour. My torch flickered and died, defeated by the water seeping through the cracks. We were left with only the feeble light of my companion's torch. The rain was icy, and my fingers clutching the compass were soon numb with cold.

As the lightning and thunder crashed around our ears I cursed the storm, which seemed determined to stop me from attaining my goal, making failure a certainty. A blast of lightning struck a rock just a few metres away with a deafening blast and an overpowering smell of electric burning. Stunned and shaken, I realised that the possibility of failure was not the biggest danger facing me. I could have died had I been just a little closer to the rock. This could have been my last day on earth.

This far graver terror might have destroyed me completely. Yet somehow that lightning strike brought a clarity and calmness to my mind. I was reminded again that my life is always in the hands of Mother Earth. I could accept that death might come to me like this, but for as long as I had life I should fight for what I needed. The powerful energy of the universe can kill you or protect you, but I had been spared this time. I was alive, and all that was stopping me from reaching my goal was my own fear of not being good enough. With this insight my mind cleared, as if the lightning had burnt away the weeds of fear that were choking my resolve. Now, instead of seeing the storm as an enemy trying to defeat me, I saw it as an ally, infusing me with the energy, determination and clarity I needed to complete the task.

And so the chocolate slab was found, the certificate obtained. More important than the certificate was that I had learnt and internalised those difficult navigation skills – and have been able to use them many times in the Drakensberg and other unfamiliar territory. But I learnt a deeper lesson as well: if I go forward with a strong heart, never losing my respect for the power of Mother Earth, I can find the inner strength and courage to face whatever challenge is before me. And always, this lesson will be linked to the crack of thunder and the haunting smell of that boulder struck by lightning.

———•———

During my first two years at the Wilderness Leadership School, I was also navigating upheavals in my family life. Some, like the birth of my first son in 2002, were very positive. Although I was no longer in a relationship with his mother, I was determined to be a good father to little Lamulani. I could not afford much, but I was committed to ensuring that he had what he needed – food, clothing and medicine.

But at this time, the turmoil in my homestead made it hard for me to focus on my studies. At the beginning of 2002, my father finally returned home after three years. He'd spent most of his retirement package, and was now relying on a government pension and a small pension from his work. He was unwell, plagued with sharp pains near his left lung. We were advised by a sangoma that these were caused by ancestors who were displeased with his behaviour, and that we needed to perform the necessary rituals to help them forgive him. My brother and I duly asked the ancestral spirits for forgiveness for my father, but we found it difficult to forgive him ourselves.

At first, I tried to push it aside. I realised that, despite my father's return, my family would continue to rely heavily on me. All my focus needed to be on trying to keep the family together and working out how to obtain the income we needed to survive.

But as my training progressed and my self-awareness grew, I began to realise that holding back my feelings was also holding back my development as a spiritual nature guide. I wanted to clear this garbage in my soul – not only my negative feelings towards my father, but whatever feelings of anger or bitterness I retained from the struggles of my childhood and young adulthood.

I had a sense of some shadowy presence that was trying to help me but was somehow stuck and unable to reach me. I went with my mother to consult a sangoma, a traditional healer who worked using only imphepho – a herb that is burnt in traditional rituals – and water to connect with ancestral spirits. He asked my mother if she'd ever lost an unborn child. She said she had – in 1973, six years before I was born. The sangoma said that this child was trying to bring light and luck to our family, and was trying to help me overcome my feelings of sadness and anger.

The sangoma gave me some purified water to sip, to make it easier for my brother to help me, and explained a simple ritual that we needed to conduct. I followed his instructions to buy a white vest, and my father performed the prescribed rituals in the family's rondavel where we connect with our ancestral spirits, including burning imphepho and slaughtering a chicken. The sangoma told me that, after performing these rituals, I should put on the vest whenever I felt this difficulty.

Whenever I put on the white vest, I could see in my mind's eye a young man, a few years older than me, waving to me from some distance. Sometimes he seemed to be on the other side of a river. I couldn't hear him speak, but he seemed to be trying to attract my attention. I spoke to the sangoma again. He said that my brother's spirit needed to come back to the homestead so that he could continue to bring light into our lives. To do this, we needed a branch of the buffalo thorn tree that was growing near our homestead.

My mother had lost the baby at a hospital in Ulundi. I borrowed a car and set off there with my parents and a friend, carrying the branch of the buffalo thorn to perform the ritual that would bring my brother's spirit back to the homestead. When we got to Ulundi, we discovered that the hospital had been demolished – but we performed the ritual at the building that had replaced it. My father spoke to the branch, which was to carry my brother's spirit. He addressed my brother by name: 'Muziwenhlahla, we have come to fetch you. Please come with us ...'

He carried on speaking to the branch, and to no one else, all the way home. He laid the branch on the back seat, between himself and my mother. After we'd stopped for some KFC everyone in the car fell asleep, and I carried on driving. When we reached home, my mother woke up and said she'd dreamt of a young man sitting in the back, between her and my father. He was sitting with an arm around each of their shoulders, hugging them. She wept tears of joy as she told us about this dream. Afterwards, we went around the homestead, showing it to the branch that was carrying my brother's spirit, so happy to have his spirit back where it belonged.

Having the sense that my older brother's spirit was there to guide and comfort me helped me to reconcile some of the difficult feelings I had for my father. I knew these were not simple things, that there were layers upon layers of complexity, that I could not make everything right in a day, or a month, or even a lifetime. But it felt as if there was a light that could guide me, as the map had guided me on that dark night in the Drakensberg, and I knew that this would help to ease my path.

I may have been facing many difficulties at home in 2002, but it was also the year in which a new love came into my life. Soon after I began my wilderness school training, I went to Richards Bay to visit my brother, who was working there. I was sitting under an avocado tree with a group of young men from my village who'd come to the town to search for work. We were sitting around a big drum on the fire in which we were cooking a cow's head – this was our Sunday routine. People walked by on their way to church, and my eye was caught by a striking young woman with a beautiful smile.

'That's Dudu,' my friend Thobani said. 'I was at school with her. She is high quality, but you'll never get her to go out with you. She just focuses on university and church. She doesn't walk up and down looking for boys.'

When she came past us on the way back from church, her parents weren't with her. I asked Thobani to tell her he wanted to discuss something with her. While they were talking, I casually walked past, and Thobani introduced me, then wandered off. I managed a few minutes of conversation with her, but she wouldn't give me her number. Later, Dudu and I would laugh at these elaborate but clumsy plans to get her to speak to me.

Some days later, Thobani managed to get Dudu's number from her younger sister. When I got back to the leadership school, I found the courage to call her. At first she was annoyed, asking me where I had got her number, but I said, 'It doesn't matter. I did it out of love, because I really want to get to know you and have you in my life.'

After a while she laughed, and we carried on talking.

I phoned her almost every day, even just for a few minutes – it was eating up my airtime, which was very expensive.

After about six months, she agreed to meet me. I'd told Paul about my interest in her – he gave me a taste of his R900 bottle of whisky to celebrate, and said I should take her on a date. This idea was new to me – I had a child already, but I had never been on a 'date'. I had no idea what to do on this date and I suggested that I take her to the KFC.

'Are you mad?' Paul said. 'You can't take a woman to the KFC! Take her to one of the family restaurants.'

After I'd been paid at the end of the month, Paul took me to Durban station and I took a taxi to Richards Bay, dressed in my red T-shirt, black jeans and shoes that a friend had given me. I took R1 000 along, and told myself that if she encouraged me to spend more than R500 it may be a sign that she was a gold digger. I met her and we took a taxi into town together. When I took out a R100 note to pay, she said, 'No, don't break your note. I'll pay. It's only R5 each.' I knew then that her interest in me wasn't the money!

We went to the Spur, a family steakhouse, and then to the cinema. Dudu was so polite and friendly to the waitresses – I would soon learn that she treats everyone with the same warmth and respect. When I got home, I still had nearly R800 left, so Dudu passed that test. But far more importantly, she had unlocked the door to my heart, and I knew that here was a woman who could really be a life partner.

Our first child, Hawelihle, was born in 2004. We were keen to get married, but it would take some years before I was able to afford this. When we did, Paul made a speech, and proudly recounted his excellent advice to me on how to manage a date successfully and win over the woman of my dreams.

———•———

After three years of training, we were no longer trainee guides but interns. Mandla and I traded our shared room at the Wilderness Leadership School headquarters for separate cottages, still on the campus but about a hundred metres from the main office. This was liberating for us – in our old room, we had always felt as if we were at work, or at boarding school. Now we could play music freely, and have visitors.

As we were now interns, we began to conduct trails, as either backup guides or trail leaders. Many of the first trails we led were with learners from wealthy schools in the cities. There are three of these in particular that I recall.

———

The promise of rain loomed in the dense, lowering sky. The hills and the valleys of the iMfolozi were blanketed with sweet grass, and the melodies of the short-clawed lark rang out in celebration of the abundance of everything. The summer rains had filled the waterholes; small streams and tributaries were gurgling with water. The soft, red soil encrusted my boots as I followed my trail leader, Mandla Buthelezi. Between us snaked a line of boys from a private school in Cape Town. They were city boys, with little experience of the bush, other than going through game reserves in vehicles. None had been immersed in the womb of the wilderness, walking for five days and sleeping in the open, with no fences or walls between them and the wild creatures that lived here.

If anyone walked on the edge of life it was Mandla Buthelezi, a brilliant and fearless guide who seemed to attract danger wherever he went. He was known as the lion whisperer, because any trail he led seemed to involve a lion encounter, but he'd had many brushes with death with other animals too. He was once knocked down by a buffalo, but he continued the trail, only going to hospital afterwards to attend to his broken ribs. He has been charged by elephants, and in a canoe capsized by a hippo. He's had two close encounters with crocodiles – one when he was drawing water from the river and a crocodile leapt out of the water as he stood back up, knocking his arm with its upper jaw, and the other when he stepped into the river without checking its depth as he could not find a stick. When his foot did not reach the bottom, he asked his colleague to pull him up. As he got back onto the bank, a crocodile come up from the depths, snapping to catch his leg just a second too late.

He attracted danger not only from animals. On trail once, I kept having the same dream every night: Mandla being kicked by black horses. When we got to base camp at the end of the trail, there were psychologists waiting for us to say that Mandla had been hijacked on the way to the park. The hijackers had taken the vehicle and beaten Mandla, but he was not seriously injured.

So walking with Mandla always made me hypervigilant, and when our path took us into the heart of the wild camphor bush that reeked of fresh black rhino dung, we walked with extreme caution.

We spotted the rhino a few metres away, grazing peacefully, and luckily unaware of us. Cattle egrets were catching a free ride on his back, harvesting the ticks around his ears. We signalled to the boys to drop their packs and sit quietly behind the tree, so that they could safely observe him. We could hear the deep rush of his breath, the crackling of twigs as he moved through the grass. We could hear an elephant some way off, breaking and uprooting a tree as he was feeding. I heard the click of dung beetles flying and colliding as they landed on the ground near the rhino midden.

I thought of my first encounter with one of these animals when I was volunteering and only a few years older than these boys. I gazed at the rhino, now so placid, but I knew all too well how dangerous they could be. I sensed in my heart that I was meeting my spirit animal, for through this animal I had broken free of my ego, and embraced the fragile edge of being that separates life from death.

Some nearby elephants trumpeted, sending a shudder of fear, and something more than fear, through the trees to the group. I could sense that these boys, who'd not seemed fearful of the rhino, were suddenly aware of their vulnerability in relation to the elephants. I could see on their faces how this awareness was bringing a sharp new understanding of the self. I recalled my own transformation after experiencing the terror of being trampled or horned by a rhino. Would their fear transform them too, bring them to a new way of being in the world? I knew little of their lives, but I knew that a privileged urban child growing up in a world negotiated through technology would have had few opportunities for connecting with wild creatures. I hoped this experience would break through those barriers, and open their minds to a deeper connection with the other living beings that share our planet.

When the rhino moved off, we picked up our packs and moved slowly away through the bush. As the trees opened out, we were greeted by a breeding herd of buffaloes spread out across the floodplain, accompanied by their egret friends. The herd moved slowly towards the river, raising small clouds of dust with their hooves. We followed them to cross the river further down, passing reedbeds alive with the twitter of weavers and southern red bishops busy building their nests. Tracks in the river sand spoke of the parade of animals that had gone before us.

As my feet sank into the warm sand of the iMfolozi River, I noticed the flowers on the riverbanks growing among the bones of old carcasses and dying trees uprooted by elephants. I passed the skull of a dead buffalo, its horns splintered by the jaws of hungry scavengers. The splintered horns were home to moth caterpillars, which feed on the keratin of the buffalo horn, then spin their cocoons in the shelter of the horns. Such is the entanglement and intermingling of life and death in nature. How delicate this fragile web of life that holds us, how deserving of our reverence and compassion. It is so easy to forget as we go about our daily lives, chasing our routines and busy schedules, numbing ourselves to the miracle of each breath. But the nearness of death in the wilderness always reminded me of this miracle.

When nature is our teacher

The trail concluded with none of the adventures that Mandla usually brought. The boys were alive with that carelessness of youth that comes from believing yourself to be untouchable, and this time the wilderness had allowed them to walk its paths peacefully. But I hoped that, in its quiet way, it had also reminded them of the miracle of life.

———————

Dark, heavy clouds swallowed us. Lightning flashed on the horizon, briefly illuminating the rain-drenched savanna. Over the roar of the thunder, we heard a male baboon barking and mothers soothing crying babies. Blinding bolts of fiery lightning ripped the clouds, as raindrops pelted the thirsty ground.

I shouted for the flysheets and my group of schoolboys came

running with them. I pitched them and we dived for cover. We kindled a fire, and watched as the umthombothi wood stubbornly burnt despite the heavy rain, releasing billows of smoke while raindrops hissed on the burning logs like angry puff adders. I could hear the mournful cries of the zebra yearlings over the drumming of the rain, and my heart went out to them. For they surely knew, as I did, that predators mainly hunt at night and are active during the wet weather when the rain masks the sound and scent of their approach.

I could see the fire dying and I warned my young charges: 'The fire must be the last thing to die!' When they'd brought more wood and blown on the embers to rekindle it, we talked about the importance of campfires in the wilderness, not only for light and warmth but also to warn animals of our presence. But the fire also brings a sacred energy, which warms us to the depth of our humanness – even a candle flame can bring this comforting energy to the night-time wilderness. For millennia, our ancestors have gathered around fires to reflect on life, to deepen their wisdom and sense of connection. The campfire has been witness to generations of ancestral wisdom being passed down through the ages, and each fire we light rekindles our ancestral connection. As we huddled against the cold and rain blowing under our flysheet, I encouraged the boys to give thanks to our ancestors who tamed this powerful force and gave us the gift of fire.

We imagined that we too were early humans, crouching around the night-time fire. I teased them by telling them that the bark of a leopard was the call of the long-extinct dinofelis, that ferocious sabre-toothed cat that was thought to prey on early hominids, with its skull-crushing jaws. They laughed, but glanced into the darkness with secret fear, shivering with the ancestral memory.

Later, with supper done, we settled down for the night. I wriggled on my back, trying to find a comfortable position. The air was filled with the scent of rain and the sounds of night creatures, the calling of the fiery-necked nightjar, the barking of a distant leopard, the fading rumbling of the thunderstorm. Gradually, these sounds hypnotised me and carried me into sleep.

I fell asleep with an awareness of the young man taking night watch by the fire, knowing I was placing my life, and the lives of all these young boys, in his hands. I wondered whether these boys had ever had this much responsibility – each boy was on duty for one hour, watching out for dangerous animals. When his turn was over, he would wake the next boy to take over. The practice of a lone night watch had been developed by the Wilderness Leadership School, which found that people are more vigilant when on their own – they are more likely to get distracted when talking to another. But I also knew from my own experience that few moments are as profound as those spent sitting alone in the darkness by a fire in the wilderness, knowing that you hold the trust of your companions. Sitting in a room full of people can be a lonely thing, but sitting alone in nature can help us to open the door to our hearts and bring healing to whatever sadness, grief or fear we find there. I was glad that these boys had been given the opportunity to experience this.

———•———

One of my most powerful experiences on the schoolboy trail was when my inner voice warned me of unseen danger, enabling us to avoid a disaster.

I woke to a still, humid morning under a hazy, dull sky. My heart was heavy with a sense of foreboding, as if a weight was crushing my chest. I shook out my sleeping mat and bag, then walked over to the fire and poked the embers with my walking stick, sending sparks shooting into the air. This told me that the burning log was common crowberry, which shoots sparks when the bone of the tree is being burnt.

The young man on morning watch greeted me, and asked if he could catch some sleep before the others awoke. I was glad to let him do so, so that I could focus on understanding the oppressive, threatening energy that hung over the place. On the flat rock shelf around me were the sleeping forms of my backup guide and the eight boys in my care, from a private school in Cape Town. The river ran below, a swirling rush of muddy water that seemed angry, but not because it was swollen by recent rains. I'd often been on trails with rivers running high from the rains, but this was a different energy, one that filled me with agitation and fear.

I moved a little way from the sleeping boys and sat meditating on this until I heard them stirring. I heard the boys talking about going down to the river to fill the water bucket for tea, and quickly rose to tell them not to go down to the water. They should tie a rope to the kettle instead, and dangle it from the high rock shelf down into the river to scoop the water up. Unfortunately, they had no idea how to make a proper knot, and when they tried to scoop the water up the kettle came loose and was quickly carried away by the fast-flowing river.

The boys wanted to get into the river to search for the kettle, and to cool off from the oppressive heat, which was baking down although it was still early. 'We're hot, Big C,' they begged (Big C was my nickname).

'Well, it's good to be hot,' I replied. 'The girls will love you.'

The boys groaned.

'Not like that, Big C! We want to swim. You let us swim yesterday.'

'Yesterday was yesterday. Today is today. And no one is going into the water today.'

I didn't want to lose the kettle, but I could not shake off this feeling of dread. I stood firm against their nagging, even when my backup guide started trying to persuade me as well. I said that they could tie our canvas water bucket onto a rope and scoop water up from the safety of the high rock instead, then pour it over themselves to cool down. This time, I tied the knot myself, showing them how to do it.

For a while things went smoothly. The boys shrieked with laughter as they dumped water over one another's heads. I moved into the shade, and began sorting my equipment and lunch for the day, still tormented by unease. I decided that we would stay where we were until I understood what was causing my feelings.

Then, 'It's stuck,' I heard one exclaim.

'Imagine we lose this bucket, after losing the kettle. That's hilarious!' declared another.

I didn't think it would be hilarious to lose another piece of equipment, especially as it was only day two of the trail and we needed that bucket. I hurried over to see what the problem was.

'It's stuck in the reeds,' the boys said.

I looked over the edge at the water, but the reeds did not move when the boys tugged the rope.

'It must be caught on something else,' I said.

One of the boys – a big rugby player nicknamed Mugabe because he was from Zimbabwe – jerked hard on the rope.

'Pull it gently, it's not a scrum,' I warned him. 'We don't want to pull the handles off the bucket.'

As I took the rope to help them, I felt a peculiar, tingling energy shooting up the rope, into my hand and up my arm. It reminded me of the time I laid my hand on the spoor of the lion. When I felt another tug on the rope, I knew what was pulling the bucket. But it was my backup guide who named it.

'Let go, Big C! It's a crocodile.'

Wayne was an old guide from Mpumalanga, who'd run his own school to train guides and was now freelancing. He had a deep knowledge of the bush, and had taught me much about geology. His words confirmed what my arm had told me, although we could see nothing in the turbid water.

But I did not let the rope go. I told myself it was because I didn't want to lose the bucket. But in my heart I knew that something more powerful was making me hold on to that rope. In my hand was the memory of that hand I had let go, all those years before – the hand of my best friend Sanele. I would not let go this time.

The force came again. It was clear that something strong and very much alive was trying to tug the bucket away from us. We tightened our grip against it. We could see nothing, but could feel the ferocious energy and strength of the animal through the vibrations and tugging on the rope. We could see movement in the water as the crocodile rolled, twisting the bucket as these animals do when drowning prey. The rope was burning and cutting into the palm of my hand, but I was determined to hold on.

The pulling stopped, so abruptly that we all fell back and tumbled over each other. We sat up, rubbing our sore hands, and pulled the rope up. The handles were there, but the bucket was gone, taken by the crocodile for reasons unknown. Later that day, we found it washed up on the bank.

Afterwards, I looked at the boys laughing and retelling their experience, talking about how they were going to boast about this to the others. I thought about what they'd told me about their ordered lives, the way every minute of their school day was bound by rules – when to eat, when to sit, or stand, or go to the bathroom, or listen to the teacher and follow instructions. The wild, beautiful beast that lives in all of us was never nourished, or given space or expression. I hoped that this experience had touched that wildness inside them; that feeling the awe-inspiring power of the crocodile had awoken some sense of the secret beast within themselves, the beast that dances in the wind and howls for the moon; and that it had brought about some subtle shift in their souls that would change them forever.

At the end of the trail, I encouraged them to give thanks for whatever spirits had protected them by warning me not to let them go into the river. Mugabe spoke about how being brushed by death had awakened in him an awareness of the razor's edge of life, and what it is to walk along it, and I was glad that he had received the gift of this knowledge.

In my heart I gave thanks to Sanele, for I had no doubt that his spirit was with me that day. I gave thanks too for the experiences in my life that had taught me to open my senses and trust the hidden voice that warns us of the unseen. I have kept the torn-off handles of that bucket. For me, they somehow symbolise Sanele's bones, the bones we never found for burial. It felt important that both the crocodile and I had come away with something, and for me the handles also represented a white flag of peace, an expression of a renewed reconciliation between my traumatised childhood self and the crocodiles, mediated by my adult self. I was grateful for this opportunity to revisit that trauma and make peace with it once more. I could not save Sanele, but by listening to my inner wisdom I had saved those boys. And that was cause for gratitude indeed.

———•———

My time with the Wilderness Leadership School opened so many new landscapes for me. One discovery that has continued to enrich my life was learning more about the value and meaning of nature in the cultural and spiritual lives of the world's indigenous peoples. While Baba Gumede had shown me much about my own people's insights into nature, I had known nothing of the belief systems of other indigenous peoples.

We had limited access to 'entertainment' at the school, but we were allowed to watch nature documentaries and a few feature films. One such film was *Dances With Wolves*, which completely captured my imagination. I had never even known that there were indigenous tribes living in North America when the settlers arrived there from Europe. I was fascinated by this insight into these peoples, and into their ways of living and being in harmony with nature.

Guided by mentors like Ian Read and Bruce Dell, the managing director of the Wilderness Leadership School, I started to do my own research about how the Native American belief systems compared to those of Zulus or other native African tribes. When I turned twenty-five, I bought myself *Bury My Heart at Wounded Knee* as a present. This searing account of the devastating impact of the European colonisation on the North American continent made a deep impression on me. Learning of these injustices drove me to read more about our own history, and the atrocities carried out under colonialism in South Africa – I was horrified to learn, for example, that until 1906 settlers were paid a bounty for killing members of the San community.

One of the documents my mentors gave me to read was the famous letter attributed to Chief Seattle, in response to a request to purchase tribal lands for mining. In this and in other readings, I was moved by the respect shown to nature, and the understanding of our profound interconnectedness. I was hungry to know more, and in time would learn all I could about ways of living with and showing respect for nature among the Australian Aborigines, the Himba people of Namibia, the Inuit in Alaska and Canada, and other indigenous peoples around the world. In addition, my mentors gave me other philosophical texts such as *The Holy Man* by Susan Trott, which explored techniques of meditation, including fasting and solitude.

I was still new to this field when we went on trail with a woman named Meredith Little. Meredith and her husband, Stephen Foster, had co-founded an organisation called Rites of Passage Inc. and later founded The School of Lost Borders. Stephen has since passed away, but Meredith is still doing their life's work to conduct rite-of-passage ceremonies in the wilderness to support people in different life stages, including those who are dying. She is also devoted to encouraging a deep connection with nature and enabling healing through nature.

When I met her, I was intrigued by this person whose face reflected a life spent in the wilderness – much of it in the harsh environment of Death Valley in the Mojave Desert, the Californian national park where she and her husband conducted their trails. Meredith was in South Africa for a conference, and had come with some of her colleagues to experience one of the school's wilderness trails. Mandla Gumede and I took turns to lead the trail, under the guidance of our mentors, who were there as backup. The trail was gentle and relaxed, with much time spent bathing in the river or walking in silence.

I remember sitting one evening under a warm blanket of cloud, listening to Meredith and our mentors speaking of vision fasts, rites of passage, practices of living and dying; of how the trails in America were conducted with a more spiritual and transformative intention, while those in South Africa seemed to be more about adventure; and of how to recreate these transformative practices on South African trails. The earth smelled deep and rich after the rains, a Goliath heron was fishing upriver, piercing the brown water with its long, sharp beak in search of prey. The damp firewood sizzled and hissed, sending sparks flying into the sky. The calling of a Burchell's coucal, its song like the gurgling of water being poured from a bottle, intertwined with her voice, harmonising like a flute accompanying vocals.

This trail was an empowering experience for Mandla and me, for our mentors gave us the space to lead and to share our own indigenous knowledge of life in the bush and indigenous culture. They were surprised at the extent of our knowledge, and listened to us seriously and respectfully without interruption, affirming that we too had wisdom to offer.

I did not understand everything that Meredith and the others spoke about, for the language and concepts were new to me, but I resonated strongly with the ideas. I already knew about the transformative power of nature deep in my heart, body and soul – I had known from the time when I wept for the sunset as a very small boy. I had grown up being seen as a strange child, who cried about the sun and fought for the lives of trees and baby rabbits. But now I was learning that there were other 'strange beings' like me all over the world. I was also learning to trust and develop my intuitive ideas, to compare them with those of others, to find a language to express them. I was emboldened by this discovery of an international community of people who felt like I did – and inspired to go further down the path of learning and embracing indigenous wisdom. The calling I had felt as a very young boy was like a spiral, coming around me to touch me again and again, but each time with more depth, with more wisdom and more power.

In 2005, I had the opportunity to conduct a trail that was very close to my heart.

We followed an old animal path on the floodplain of the Black iMfolozi River, walking beside wide, round elephant tracks. Brown scrub robins sang in the bushes nearby, and the air hung heavy with the scent of dry grass and wild basil. The heat from the barrel of my firearm coated my hands and arms in sweat (we always carried our weapons ready to be fired, if necessary), while small biting flies tormented my face, ignoring my efforts to bat them away with my left hand.

The path took us straight to the steep Siwasomsane cliff, and I knew we had a tough climb ahead. But I walked with a light heart, for one of my greatest dreams was starting to manifest itself.

I had now been with the Wilderness Leadership School for over three years. I had been on or led many trails with teenagers from private schools in Cape Town and Johannesburg, and with wealthy white people from Europe and South Africa. This trail was different. The Wilderness Leadership School had been approached by the Thembeni Crisis Centre, which worked with at-risk youth and wished to run environmental clean-up projects in Durban. The centre wanted youngsters to experience an unpolluted environment, that they might inspire others in their community to keep their streets clean, and had found sponsorship to do a trail. The CEO had selected seven youngsters for the trail, and one employee, an older man, to accompany them. I had met them in Durban previously to prepare them for what they might find in the wilderness and guide them on how to behave: it would be their first time in a wilderness area.

I paused on the path to look back at the faces of the young men following me. I recalled the young boys I'd come across as a volunteer, peering through the fence to see the animals. I deeply understood their hunger to discover the wild creatures who shared our world, their hunger to embrace the sacred ground that had been ripped from their forefathers, the ground to which they were still denied access.

I recalled my conversation with the young poacher, who'd said that his main reason for joining the poaching trip was to experience the animals; how, among my age mates, I alone had been privileged enough to visit the iMfolozi because my father had worked there. I remembered my longing, as a young child on the banks of the iMfolozi River, to find a way to bring this experience of the wilderness to my community.

Now, at last, I had my first chance to lead a trail with black people – and my first chance to lead a trail in my mother tongue. How liberating this was, to be able to express my thoughts freely, not to have to explain my cultural references. Although my English was improving rapidly, I still found it tough to conduct trails with non-English speakers from Europe, as I struggled with their accents and they struggled with mine.

I turned and continued the steep climb. My muscles complained as the weight of my backpack pulled on my shoulders, the grasses scratched my itchy legs. I tried to embrace these discomforts, as focusing on them can hinder the journey of the spirit. Behind me, the harsh breathing of the elder from the Thembeni centre reminded me to focus on my own breathing. This helped to push distracting thoughts from my mind. I was reminded again about how nature always gives us the choice: to focus on the discomfort in your body or the anxieties in your soul, or to be fully present.

I felt myself come alive to the beat of my heart, the thump of my boots on the ground, the soothing, cool breeze on my arms, the splash of water as an elephant drank nearby. My body became light, my spirits soared to fly with the bateleur riding the thermals above us. I walked towards the cliff edge, carried by the lightness and life flowing through me, at one again with the world inside and outside of me.

As we came to the edge of the cliff, we were rewarded by a view of the snaking brown waters of the iMfolozi. My binoculars showed crocodiles sunning themselves on the riverbanks, looking as harmless as scattered driftwood from that distance. A group of buffalo bachelors lay on the sand; various other creatures wound down the ancient paths to the water. The young boys threw down their packs, and exclaimed in wonder at the sight that stretched out before them.

Born and raised in urban townships, they had never walked in the wilderness or slept under a clear night sky vibrant with stars. They were leading the harsh lives of township dwellers, with no time to discover themselves, suffering the cruelties brought by poverty and other life challenges. But here, at least, there was space for them to breathe, and the silence to hear themselves breathe; here, they could walk the paths of their forefathers and rekindle their vanishing culture, heal old wounds of displacement and dispossession, ground themselves and discover their inner wild animal.

Seeing these youths come alive to this wild space was like watching my dream come true – my dream to help others find their own dreams by walking these wild paths. I knew that opening the wilderness experience to my people would be a tough climb, far tougher than the climb we had just done to reach the cliff top. But today I was here, with these boys, and had woken up to find myself living the stories in my dreams.

———————

In the early spring of 2005, I was given the opportunity to lead a trail whose participants sought a spiritual experience rather than a bush adventure. This experience showed me again how the energy of a group is so often reflected in the living world.

One of the most powerful aspects of Wilderness Leadership School trails is walking in silence. This is critical to ensure that you remain vigilant, and also to ensure that you don't frighten the animals away with your chatter. But it is also a powerful experience in itself.

Being in silence is a particularly powerful and much-needed medicine in a world that has become so contaminated with noise that bird species are threatened because they can no longer hear one another call, and many humans are driven to mental illness.

While all the groups walk without speaking, this trail was with a group of women from Europe who specifically wanted to conduct most of the trail in silence, using it as a way of reconnecting with their inner selves. We set out after the first rains in mid-August, a time of softening in the landscape as flowers and green grass start to sprout, and the unhloyiya, the yellow-billed kites, return from their migration. These women brought such gentle, loving energy to the wilderness, embracing everything before them, the flowers, the birds, the tiniest insects, the shape of the trees. I was moved by the energy and attention they gave to everything, by their compassion and passionate yearning for change, for a better future for all living things. Walking with this group reminded me how much the wild creatures mirror our intention.

This group radiated a gentleness that rippled out and radiated back to us. I found myself stepping lightly on the path, careful to avoid the delicate petals of the new flowers. The air was heavy with the sweet scent of the white and cream flowers of the acacia trees on the hills. Giraffes glided past, nibbling the new leaves, wrapped in their own dreamlike quietness. We lowered our packs for a break near a group of warthog kneeling to graze on fresh grass shoots. As we sat quietly, our own inner silence enabled us to hear, sharply, the quality of each sound – the gulping and gurgling of water when someone drank from their bottle, the plaintive call of the trumpeter hornbill in the fig tree, so like a baby crying for love. When we speak of being in silence, we mean silence from human chatter (with one another and inside our heads). The bush is seldom silent. But the sounds of nature at peace with itself – the whispering wind, the rippling of the water, the hum and chorus of birds and insects going about their lives in the trees – could always bring me to calmness.

We continued walking, up a path that slowly ascended the surrounding hills. We passed dung and tracks, skirted the trunks pushed over the path by elephants, stopped to consider the rubbing posts, their smooth shining surfaces smeared with mud or ticks or blood. The hills undulated around us, some broken by sandstone cliffs, others graced by giant umbrella thorns or magnificent marula trees; kudus and sunbirds nibbled on weeping boer bean flowers.

The path took us across a stream, fringed by the hanging branches of a small umthombothi tree. We took our boots off, put our feet in the clear water and sat listening to the soft chirruping of the brightly coloured gorgeous bushshrike in the thicket. We were united in our gratitude for the moment.

Later, we sat by the river at sunset, watching the water turn gold. Impala, wildebeest and buffalo herds drifted down the grassy hills to drink, then a rhino female with her playful calf, charging and splashing. An elephant bull rumbled softly in the reeds. Darkness fell around us as we sat by the flickering fire, bringing the night sounds, the low whooping of hyena by the river, the hooting of a Verreaux's eagle-owl nearby. The moon rose, full and bright, dimming the nearby stars so that it seemed the stars were shying away from its brightness. We sat wrapped in silence, our minds and senses empty of the noisy clatter of modern life, our hearts wide open to the harmonies and rhythms of everything around us.

——————

In the following winter, two years after listening to Meredith Little, I had the opportunity to lead a rite-of-passage trail myself.

It was deep in the seasonal winter drought when we set off. The white grass and dry leaves swirled in the wind as a crested francolin darted across our path. A dusty haze hung over the hills, and the river was reduced to a thread of puddles in the pale sand. A strong odour alerted us to fresh dung in a rhino midden, as two dung beetles hauled their dung ball ponderously alongside our path. The wind rattled the dry branches of the naked wild camphor bush as we walked towards the river. On the far bank, a breeding herd of impalas nibbled fresh reed shoots, while nyalas, heads lowered, fed on fallen leaves.

We sat down and took in the moment in silence, looking out at the sandstone cliffs beyond the river, the hillsides with their spreading umbrella thorns, basking in the warmth of the winter sun. As we took off our boots and prepared to cross the iMfolozi, I said to our companions, 'When you cross the iMfolozi, you are symbolically crossing into the wilderness. The water will wash your feet clean of the city dust. On the other side, your feet will be coated with the pure dust of the wilderness.'

But Tegan, one of the young men on the trail, had come to the iMfolozi wilderness for another symbolic crossing. His father, Richard Knight, had brought him on the trail for a rite of passage from youth into adulthood. He wanted to give Tegan an opportunity to dance with his inner and outer wilderness. Like an elephant bull teaching a younger bull the migration routes across the desert to lush pastures, Richard was giving his son the simple but profound blessing of a father's support – and a father's permission to discover his own true self.

In many societies, the rite of passage into adulthood has become an occasion for lavish parties with alcohol and expensive gifts, one that offers little to guide the young adult through life or give them a sense of purpose or identity. My reading into indigenous cultural practices had reminded me of the importance of the many rites of passage that we have lost altogether, or that have lost their true meaning. Young adults are seldom empowered with the wisdom to live in harmony with one another and with the natural world for a good understanding of themselves – leading to many of the problems we face today. The adult world can be dangerous and hostile, and many youngsters end up depressed and anxious, making unsuitable life choices. A good rite of passage can help to root young adults, helping them to see more clearly what should guide their choices.

So this was an exceptional gift that Richard was giving his son: acknowledging the significance of this transition by bringing him to this place of profound crossings, rather than letting him stumble into adulthood. And what better place to cross into adulthood than this wilderness? The bright, warm days, the haunting call of the African fish eagle, the bright flash of a lilac-breasted roller against the sepia winter foliage ... each experience was a unique treasure, a reminder of the limitless beauty of our world. The solitary hour spent by the fire on night watch offered such an expansive space for contemplation ... all these experiences could help to initiate an adult.

Tegan was twenty-three – the same age as I was at the time – and the two of us became good friends. He was from Scotland, his father a CEO in an upmarket whisky company. I was from a rural village in South Africa, the son of a man who cared for horses. But we found much to share, and spent long hours talking about what it was to be on the threshold of adult life. I had the advantage of having gone through an initiation ceremony. I spoke to Tegan about what that had meant for me, about everything uMkhulu had taught us about how to walk through the world as a man of peace, taking responsibility for all who need your care. Richard also appreciated what I was giving his son, and would himself later become an important figure in my life.

This was my first trail that had a conscious intention of being a rite of passage. In later years, I would guide many trails on which a father and son or mother and daughter came into the wilderness to cross the threshold. We often had experiences that would resonate with this intention. On one trail, I watched a white-backed vulture flying with its chicks for the first time (I knew it was the first time as I had been doing back-to-back trails and had been watching that vulture nest). Another time, we came across a lioness teaching her cub how to kill a small warthog baby. It was as if the wilderness was supporting the initiations by giving us these messages.

I mentioned earlier that the Black iMfolozi was called the Imfolozi Emnyama inkethabaweli – the river that chooses who crosses it. The river was 'black' because of the slate stones upstream, but also for its dark, coffee-coloured waters that concealed all manner of dangers – crocodiles, water snakes, hippos. For many, crossing the threshold into adulthood can be a dangerous time. Young Xhosa initiates go into the wilderness after circumcision; some do not return. Young people take risks and

Crossing into the sacred space of the wilderness

expose themselves to many dangers before they acquire the wisdom of age; young men are sent into battle. But it is also an emotionally dangerous and confusing time. I was so pleased that I too could help to lead Tegan across this threshold – and the many young adults who have since made this crossing with me in the iMfolozi wilderness. They came from all over the world, and had grown up in circumstances very different from my own. But through them I learnt that people at the same stage of life face some of the same challenges. And helping to usher these youths into adulthood enabled me, each time, to reflect on my own journey.

●────────●

Soon after the trail with Tegan, I had the opportunity to deepen my practice as a spiritual guide when I conducted a trail organised by a tour operator who had travelled in the US with a Native

American shaman. He wanted to bring some of the shamanistic ways of connecting to nature into his tours. Thanks to my encounter with Meredith Little, and the reading I had done, these were not new to me. Some even echoed our own traditional rituals in Zulu culture.

The group was made up of young professionals from England, most of whom worked for a big chartered accountancy firm. They'd left busy corporate lives in London, with punishing schedules of late-night parties, meetings and deadlines, for a wilderness experience to revive their burnt-out hearts, bodies and souls.

The trail began with smudging – the Native American practice of bathing in the smoke of a bundle of burning sage as a form of spiritual cleansing. The smouldering bundle is passed over your body, under your legs and armpits, and near to your face. The concept was not unfamiliar – in our local indigenous cultures, imphepho is burnt as a cleansing ritual, although we don't do smudging in the same way. The tour operator had brought wild sage from America, and I contributed imphepho and red ochre to connect the trailists with our earliest ancestors from Africa.

We didn't do any other formal rituals, but because the group wanted to experience the trail in a more spiritual way we were able to reflect on our experiences each day more deeply. The tour operator was an older, calm and easy-going person, who was happy to step back and let me take the lead. This was a great opportunity, for it gave me the space to explore ideas. In the mornings, I would invite the others to reflect on their 'inner weather', using weather as a metaphor for their internal mood. Was it stormy, calm and sunny, windy and turbulent, cloudy and uncertain? In the afternoon, people borrowed features from the landscape around them to describe their state of mind. They would say, 'I am sitting in a big tree covered with flowers, looking out over the hills,' or 'I am sitting on the edge of the red donga, with my legs dangling, and the ground is falling steeply away below my feet,' or 'I am canoeing on a calm, still river, with fish eagles calling above me,' or 'I am lost in a thicket of thorn trees.' This exercise deepened the trailists' awareness of the outer landscape and weather, while giving them the vocabulary to express the subtle nuances of their inner landscapes.

On our third night, the dark, swirling clouds blanketing the sky were cracked open by blinding forks of lightning. Rhinos drinking from the river jostled and squealed in fear; a ghostly wind shook the branches above us, scattering us with leaves. I hurried to my backpack and scooped up my flysheet as the heavy drops started to fall.

As the rain started bucketing down, I checked that the sleeping bags were out of the rain. We'd had rain every evening, and the group were familiar with the routines of covering the sleeping bags and vulnerable equipment, and protecting the fire by covering it with a kettle of water.

Once the flysheets were up we sat by the hissing fire, grateful for the umthombothi wood, which keeps burning even when it rains. The water of the iMfolozi changed from silver to the colour of weak coffee. The tracks of the rhino and other animals were washed away, leaving the sand on its banks smooth and clean.

As the downpour eased to a gentle fall, I heard the group complaining about the weather. They were upset: they'd assumed they'd be spending every night under the stars. They thought that it was always warm in Africa, that the rain never came.

You cannot travel a path until you become the path, as the saying goes. These trailists wanted to travel the path of the wilderness. But they weren't willing to be the path of the wilderness. Being the path of the wilderness means being as one with whatever the wilderness offers you. If you are at one with the path, you will not be challenged by what the path brings.

We are used to manipulating our environment for our comfort and convenience. Even so-called wilderness experiences are often packaged to tourists in a way that suggests that they can 'order' what they want – an African sunset, a lion kill, the Big Five – as if they were ordering Coke and chips at McDonald's. I once had a tourist say to me, 'Ranger, tomorrow morning I want to see two leopards mating.' 'Good rangers' are the ones who can find the animals the tourists have ordered.

These chartered accountants had come seeking a true wilderness experience, but still felt that they should be able to order one. They saw the rain and clouds as obstacles depriving them of a perfect experience. Because they were blaming the wilderness for inflicting weather that they didn't want, they were closing themselves to the beauty, power and wisdom that the storm could offer.

To have a true wilderness experience, you need to be present to whatever the wilderness is offering, whether you see a lion or a grasshopper, whether you walk through sunlight and gentle breezes or howling winds and driving rain. Embracing whatever the wilderness offers opens you to oneness with all that is around you. When you are one with the flow, you become the flow itself. You do not just travel the path – you become the path.

As we spoke about this, some of them began to think about the rain differently and I could see that the lightning was burning away the roots of their expectations, of that disconnection with our surroundings that comes with the choices, comforts and convenience of modern civilisation. The thunder took my words and echoed them through the hills. They wove into the liquid warbling of the Burchell's coucal celebrating the rain, with the dancing and sighing reeds. We sat in awe of the transformative power of nature.

That night the stars came out, and I urged the trailists to bring their sleeping bags out from under the flysheet to experience this wonder. I unzipped my own sleeping bag to lie, relishing the cool breeze on my skin, beneath the constellations of Scorpio and the Coalsack. The night sky was studded with stars far beyond the counting, some shooting across the heavens in a dying blaze of glory. I lay drifting in that magical land between sleeping and waking, weaving the sounds around me into my dreams – the crackling fire, the catfish splashing as they jumped for insects, the yowling of the hyenas trying to catch the leaping fish. I noticed the excitement of the young man doing the night watch, who jumped up for a closer look. He shone his torch into the river, starting when it caught the reflection of two hyena eyes staring back, causing him to spill his tea, burn his fingers and drop his cup. All this happened without his realising that I was watching him.

In the early morning the river was riding high, swollen by heavy rains upstream. Waves splashed against the fragile banks, logs and other debris rushed past, carried by the powerful currents. A hamerkop flew above the shimmering mist hanging over the water; a group of waterbuck gathered on the far bank to drink, the lead male's formidable horns emerging from the pale mist like those of a mythical beast. Around me, the group was beginning to wake to this African dawn, to the mingling smells of coffee and river mud. The morning was blanketed in stillness, devoid of the bird song that greets a sunnier day. I watched the trailists emerge from their dreams, and hoped that today they would be able to let go of their expectations, their busyness and stress, and become one with whatever nature would offer us. That today, for a short time at least, they would become their path.

———

The Wilderness Leadership School trails attract people from all over the world, who come seeking healing and solace. I already knew that being in the wilderness was my most powerful medicine. But it was so moving and inspiring to see how the wilderness could reach others who were hurting. Every day, the wilderness showed me some new expression of its healing powers.

Like all humans, everyone who comes on the trails – even the young schoolboy – carries some sadness, and just by being in the wilderness everyone who comes on the trails will experience some easing of their sorrows. But the Wilderness Leadership School also caters for people who have suffered extreme trauma or conflict, who come on a trail with the specific purpose of finding healing and forgiveness. The Glencree Centre for Peace and Reconciliation is an international organisation aimed at promoting peace in Northern Ireland and other conflict zones. It sent many groups on trails as part of its reconciliation programme.

My first trail with the centre was in 2006. There were people from different sides of the conflict in Northern Ireland, and all had suffered trauma – Protestants and Catholics, people from the British military and from the Irish Republican Army (IRA). From the beginning, I could sense the tension in the group and their hunger for reconciliation. This was so different from a trail for schoolboys looking for adventure. These people were hurting, and I could sense the subtle energies of the wilderness shift around them. As we walked on through the dry winter landscape, past the branches of the naked trees, I could already perceive a slight easing in the group. Stepping into the wilderness is like stepping into a cocoon of transformation – you don't know how you will be changed by the experience, but you will be changed.

The experience and structure of the trail itself brought about significant shifts in the dynamics between the group members. Former enemies were now relying on one another for help to cross rivers, to get their packs onto their backs, to gather wood and cook. As we settled down for the night, an IRA cadre who had been interrogated and imprisoned by the British military was undertaking to keep the military officers safe by doing his shift of the night watch. His one-time jailers, in turn, would be getting up later to keep him safe.

But as we walked on, it seemed as if the wilderness community was also harmonising itself to the needs of the group and offering simple but profound opportunities for healing and reflection. On the second day, I heard inkwambazane – the emerald-spotted wood dove – calling from the depths of a nearby thicket. This little dove has the grey plumage that most doves have, but is distinguished by two brilliant flashes of emerald on each wing. It has a fluting call resembling a panpipe or woodwind, which slows down and drops off towards the end, fading like a lament.

Instead of identifying the bird by its scientific name, I invited the trailists to listen to the different bird calls they could hear and comment on the emotion or mood they expressed. They all noted the sadness of the wood dove's call, compared to the yellow canaries' cheery whistling or the strident bellow of the hadeda ibis. When I told them the dove's call is interpreted by the Zulu as crying, 'My father is dead, my mother is dead, and my heart goes totototo,' one of them exclaimed softly, 'Oh, my God, this is beautiful. It is talking about us, about our people killing each other.'

As we walked on, the emerald-spotted wood dove kept pace with us for a long time, flying ahead and uttering its call until we went past, and then flying ahead again. This behaviour is common in a honeyguide, which might follow you all the way home then keep tapping on your kitchen window to attract your attention. But I had never seen such behaviour by a wood dove. It was as if it did indeed sense the burden of silent grief being carried by these people, and was giving voice to that sadness.

We walked on, under the simple blessing of the sunlight on our skin and the intermittent shade of the umthombothi trees. On the ground beneath the trees, their seeds danced and cracked like popcorn kernels on a hot stove. I explained that the noise and movement came from caterpillars inside the seeds trying to break out. These people are also needing to break out, I thought. To break through the shell of their anxiety and grief, and step into the light of forgiveness and reconciliation.

On the following day, the group was sitting reflecting on some of the traumas they'd suffered. A herd of impalas was nearby, browsing peacefully on the small bushes around. A leopard approached them, but they carried on browsing, seemingly unconcerned. The group asked me why the impalas were not fleeing at the sight of the leopard. I explained that the leopard was not hunting – it was walking calmly, not stalking. I could tell from its body language that it was not hunting, and so could the impalas. This led to a discussion about how an impala could make peace with a creature that would usually be its enemy, how it could judge the leopard by its intentions, not by who it was. And why it was so difficult for humans do the same.

On our last day, as we crossed the river, we saw a lonely buffalo lying peacefully on the sand beside the iMfolozi River, twitching its ears. On its back were three red-billed oxpeckers, hopping all over, searching for ticks and other parasites. I explained what the birds were doing, helping the buffalo by removing ticks and also warning it about approaching predators, just as the buffalo was helping them. This led to a discussion about ubuntu, how much humans need one another – especially the help and comfort of others in times of sorrow. Witnessing this simple act of mutual kindness helped the group foster a reconnection with themselves.

People seldom spoke directly about the hurt they had suffered, but one afternoon a young woman was washing pots with me in the river. As we sat in the shallows scrubbing the pots with river sand, she told me that she had lost her brother and parents when a bridge they were driving over on their way back from church had been blown up by an IRA bomb. She said that coming into the wilderness and walking on African soil had calmed her soul. She felt at peace because she could strongly feel the spirit of her parents in the iMfolozi, which helped her to start thinking that she could forgive the people who had killed them.

As the trail days passed, I reflected much about the hurt these people had suffered, the wounds that so many humans carry. I thought about the weight of sorrow, how it can pull us down and make us feel lost and confused. I knew from my own traumas that forgiveness is the most critical step in healing. Not forgiveness in the sense of exonerating the person who has done you wrong, but in the sense of laying down your burden of sorrow. Forgiveness frees you from the prison of regret, releases you from blame and anger. It enables you to focus instead on finding the strength you need to step into the unique and singular purpose for which you were created.

I thought of all we had encountered: how the wood dove had given voice to the group's grieving; how the larvae trapped in the umthombothi seeds had reminded them to keep knocking on the hard shell of their sorrow and anger; how the impalas had shown how to be at peace today with the enemy of yesterday; how the oxpeckers and buffalo had affirmed the simple but profound principles of ubuntu and collaboration. The wilderness had offered all these things, but also just space and silence, which had allowed that young woman to open herself to the spirits of the ones she was grieving for and lay down her burden.

I know that finding the state of grace that comes with forgiveness is not simple. It is especially complex if someone close to us has been hurt or killed. We might feel that we are letting them down if we let our anger go. The wilderness gives us space for grief's work to happen, sometimes without our even being aware of it. Until we realise, one day, that we can let our sorrow go, as lightly as opening our hands to release a silver milkweed seed and watch it float away into the blueness of the sky.

———•———

I was learning so much at the Wilderness Leadership School. But I also knew that I needed to deepen the indigenous knowledge that had been passed on to me by my elders and ancestors. Exploring the indigenous wisdom of other cultures was also offering me a new way to explore my own. And one plant that holds much indigenous wisdom is uMlahlankosi, the buffalo thorn.

At each node on the buffalo thorn branch, there are two thorns: one curves back towards the roots of the tree, where the branch has grown from; the other points forward, to where the tree is going. This echoes the Zulu idiom 'impumelelo yakho ayingakwenzi ukhohlwe yimuva lakho', which means 'do not let your success make you forget your background' – embrace your roots and the people who supported you during your development.

My training at the Wilderness Leadership School took place in that tumultuous decade after South Africa's first democratic election in 1994. My country was hungry for transformation. I believed I too had a role to play in this, although I was unsure of how I could best help my community. My training was opening many paths for me. But I knew I should never forget the wisdom of the buffalo thorn – that, as much as I should look forward, I should also look back.

As new opportunities opened up for black people, many, in their haste to put on the clothing of success, were quick to cast off or look down on the old ways. But the more I progressed, the more I began to appreciate my origins.

The rural community that raised me had in some ways been less shaken by the white oppressors, hidden as it had been from the troubles of the country. But we could not escape many of the harsh legacies of colonialism and apartheid – those who left to work in the city were subjected to cruel and arbitrary laws, and at home we endured many hardships. And, as I have already mentioned, acquiring an education was a brutal affair for rural children.

But we endured the emotional and physical wounds of apartheid because of our rich cultural traditions, our social cohesion and the abundance that comes with living in harmony with nature. The fertile soils along the iMpelanyane eMnyama

River produced sufficient crops, including maize and sorghum. Sorghum, an ancient indigenous grain, was well suited to our conditions. It gave us a palatable, nutritious porridge throughout the dry winters, as well as our traditional beer, umqombothi, when fermented. We had cattle and other livestock for meat and fresh milk. Our community life was founded on principles of ubuntu and sharing. At harvest time, the community would work together to harvest for one family after another until the whole community's harvest had been collected. The rivers, soil, rain and sunshine were cherished as the bringers of life.

Traditional healers and shamanistic izangoma helped to keep the community free of illness. There was a wealth of knowledge about the healing herbs and plants that grew in the area – these chemical-free remedies worked naturally with the body and produced few side effects. The rites of passage for young people, and the responsibilities they carried from a young age, helped them to ground themselves, giving them a strong sense of identity and purpose, leadership qualities and a willingness to work for the good of others.

Being grounded in traditional values did not prevent progress and change. Despite its many hurdles, our community greatly valued education, recognising schoolchildren as the engines that would bring light and blessings. People hoped that education would bring the changes that would enrich all lives in rural communities.

So it was that I grew up steeped in indigenous wisdom – not stuffy, outdated traditions, but a dynamic world view that was rooted in ancient connections to the land and to others, revitalised as each generation infused it with new life. This wisdom gave us a way to understand our connections to one another, to our bodies, to our ancestors, to our fellow creatures, and to the earth.

In my heart, I knew the power of this perspective. But when I began my journey as a volunteer, and then as a trainee and intern, this way of knowing was not highly valued. Instead, western scientific knowledge – botany, zoology, ecology, lessons acquired from textbooks and guidebooks – was emphasised. I imbibed this knowledge and could regurgitate it, but my indigenous knowledge gave me a richer system to draw from. Indigenous knowledge is not easily learnt. It is subtle and elusive, and grows in and around you slowly, coming to you with quietness and depth, feeding your heart and soul as much as it feeds your brain. There is a danger, then, of modernisation eroding it more quickly than it can grow.

There are many ways to know a thing, and the buffalo thorn has a lesson for us here too. I know from my book knowledge that the buffalo thorn, *Ziziphus mucronata*, is a species of tree in the family Rhamnaceae, native to southern Africa. It can reach a height of up to seventeen metres, and thrives in a wide range of habitats and soil types, favouring termite mounds and soils that rivers have deposited. I know from my bush knowledge that, despite its thorns, this tree is the beloved food of many browsers. At our homestead it was eaten by goats and sheep, but in game parks it attracts black rhinos, giraffes, kudus and bushbuck. Its bark is a popular food for aardvarks, with humans, monkeys, goats and baboons eating its fruits. I know from my childhood that traditional healers use the leaves to cure boils and infected wounds.

But from the wellspring of my people's indigenous wisdom, I learnt that the buffalo thorn is also uMlahlankosi, a tree of deep spiritual relevance to the Zulu people. It got its name, which means 'the tree of the tribal leaders', from being used as a grave marker for tribal chiefs. But it has another very sacred function: it is used to gather the spirit of one who has died far from home, and to bring that spirit back to the homestead. In a culture where connection to our ancestors is central to our spiritual, mental and emotional well-being, it is vital for us to know that the spirits of our ancestors have found their way back to our homestead. This was the ritual my family used to bring back the spirit of my older brother.

The relatives of the deceased lay a branch of the tree on the spot where that person died and call the deceased person by name three times. They will say, 'Luyanda (or whatever the person's name is), you do not belong here in hospital (or wherever the person died). We are here to collect your spirit. Please come along with us, for we are here to carry you home.'

Then they take the branch home. The person carrying the branch may not talk to anyone else on the journey, as they are communicating with the spirit. If they are using public transport, they must pay a fare or buy a ticket for the branch. If they need to cross a river and lay the branch down to take off their shoes, for example, they need to explain this to the branch: 'This is not home. We are now crossing the Black iMfolozi River – we are still heading home. Please, Luyanda, come with us …'

When they arrive at the homestead, a goat may be slaughtered to welcome the spirit. Some people lay the branch on that person's grave, if the grave is there; others keep it in the special rondavel where they connect with the ancestors.

So when we look at a buffalo thorn tree through the eyes of deep indigenous wisdom, we don't just see a tree – we see a living being that brings comfort in times of grief, protects us from the distress of a wandering spirit, and helps us to keep our families together and whole.

My journey of discovery into the ways of the Native Americans and other indigenous peoples helped to affirm the value of my own indigenous knowledge. And my mentors, Bruce Dell among them, taught us not to be scared of a white skin and encouraged us to understand that we all have the right to be equally respected – we should not fear to speak our truth to anyone. They helped to lift the layer of inferiority that had been imposed on those of us who grew up in the rural areas under apartheid, which made us hesitant to disagree with a white person even if we felt they were wrong. Nevertheless, the knowledge systems of my people had been pushed into the shadows by so many colonial and apartheid regimes that it was hard to overcome my doubts about their value for other people, particularly people from Europe.

Early in my internship, I met a woman from Ireland who helped me to realise that my indigenous knowledge offered something of great value to others; and that, because I possessed this knowledge and wisdom, I could bring something unique and transforming to people's lives.

I first met Sarah (not her real name) on one of the trails organised by the Glencree Centre for Peace and Reconciliation. I was the backup guide on three Glencree trails in which she took part. She took an interest in me, and started referring to me as her 'son'.

On the third trail, I explained the spiritual significance of the buffalo thorn to her. She was very moved by this account. 'It's so important that your community doesn't lose these sacred practices and rituals,' she said. 'So many cultures have lost them because of urbanisation and modernisation, or because people's cultural practices have been marginalised or disrespected.'

She began to tease me, to test my understanding of her Irish accent – because, she said, on her next trail she wanted me to be the leader. 'If I book a trail I want you to lead it, so you have a duty to learn quickly from your white mentors,' she said.

Up to that point, I had not led any of these more challenging trails, such as those organised by the Glencree Centre – trails on which we would be expected to hold our own in discussions about philosophy and global issues. Our white mentors led these trails. So when Sarah said this I laughed as I thought she was joking.

A year later she booked a personal trail, and asked for me to be trail leader, as she wanted to hear more about the history and ecology of the park from a Zulu perspective. My mentors were willing to step back and let me lead, with Ian Read as the backup, as they were keen to develop my skills as a trail leader.

I felt like a snake shedding its old skin for a new one of authority

With Ian Read

and leadership. It was intimidating, but I knew I had Ian's support, and that he would help where needed without making me feel like he was waiting for me to make mistakes.

I stepped into this new role and felt my confidence grow as I spoke about wilderness philosophy. People listened. When we took a break, I overheard a trailist saying, 'It tastes better walking behind an African guide who was born just next to this park.'

I considered his words as we walked among the umbrella thorns, watching a swarm of bees being followed by a hungry honeyguide. I looked up at the spreading canopies, and remembered how I and my friends had fought for the life of the umbrella thorn that had been cut down so many years before. Just like that tree, my people had been cut down again and again.

But just as the seedlings of that broken umbrella thorn had grown, so we as a people were growing, reaching new branches into our democracy. This trail affirmed not only my skills and talents as an individual, but also the value of the indigenous knowledge and belief systems of my people. It helped me to celebrate my history and culture, reinvigorated my pride and inspired me to protect and share my spirituality and traditions. It affirmed the wisdom of the buffalo thorn, for as much as I was moving forward, I was still growing in my appreciation and understanding of where I had come from.

———•———

The sun was beating down, turning the pale complexion of the students from England a deep pink. Young Burchell's zebra males were sandbathing under the umbrella thorns in a cloud of red dust, their plaintive whooping calls echoing across the valley. A wildebeest herd stared at us and snorted, as if in disbelief that we were mad enough to walk in such heat. The air pressed down on us as heavy clouds gathered, promising rain and thunderstorms later in the day. My iMfolozi was crying out for this rain. Dust swirled into the sky as a herd of buffaloes pounded along the riverbed, searching for puddles.

This was my last trail with the Wilderness Leadership School, and the baking landscape awaiting the rain seemed to echo my mixed emotions of sadness and anticipation. Like the young male zebras, I was lamenting my departure from my colleagues, my mentors and an organisation that had transformed my life. But I was also thirsty for change.

During the five years that I had spent at the school, the world had seemed to open like a galaxy around me, an ever-expanding universe of discovery in which I was no longer separate from nature but becoming more and more entangled in it. I could feel myself growing into one who walks, sees, touches and smells as a creature of the wilderness. Each new wonder I encountered expanded my capacity to learn, each new learning expanded my comprehension of the power and magic of nature. I had also developed new and deeper intelligences to enable me to experience the power of the wilderness through my journeys of discovery into indigenous wisdom. And I had learnt to locate myself and my wilderness in the complex matrix of contemporary global issues.

But things had changed in my personal life too.

I was eager to get married – our son, Hawelihle, was over two years old, and Dudu had waited long enough for us to settle down as a family. I knew that together we would make a strong partnership. She was so humble, so wise, so supportive of me, and so strong. I was looking forward to building our lives together; to walking with her and our children in the wilderness and drinking from the same sacred springs, and weaving a strong foundation for our son and any other children to come, as the weavers weave a nest for their young.

But my dream of getting married was impossible under our employment conditions and pay structure, which also did not offer the security of a pension or medical aid and long-term employment. I wanted to be able to give my children opportunities I'd never had, including the chance to attend good schools and university. My ancestral homestead also needed my support – my mother and brother and were still living in rickety huts, and I was keen to build brick dwellings for them. Dudu understood why this was important for me, that I needed to build a foundation in my homestead through which my family and my children could foster our ancestral connections. These needs were pushing our dream of getting married further away, especially with my limited income.

So I was excited when a friend sent me an advertisement for a trails officer post with Ezemvelo KZN Wildlife – the organisation needed a qualified wilderness guide to conduct trails at iMfolozi. It seemed like the job was made for me, and I applied. I was keen for the post, but when I received the phone call to say that my application had been successful, beating those of more than fifteen candidates, I wept. For I realised that now I truly would be leaving the Wilderness Leadership School.

I thought of everything the school had offered me: the skills, experience and wisdom to become a respected wilderness guide, already known among those who had been on my trails; and the opportunity to walk by the sacred lakes of the iSimangaliso World Heritage Site, to sleep in the caves of the Drakensberg, and to follow the wild paths of the iMfolozi wilderness. The school had opened the way for me to find myself in the wilderness, and I had deep respect for the organisation. I knew there might be opportunities for growth in the organisation, with the promise of trails in many different areas. But I also knew that my reasons for leaving were sound. Difficult as it was, this was the path I needed to take.

When I had sent my letter of resignation, I was called to the founder's office and met Ian Player for the first time. The wind was blowing hard, setting the bamboo behind the office squealing and dancing. The roof rattled and creaked; clouds scudded across the sky and swallowed the sun. The trail equipment I'd helped my colleagues to clean swayed and sighed in the wind. It seemed the winds of change were blowing indeed, pushing me to a new phase of my life.

I sat on a chair in front of Ian Player, drinking the tea I had been offered. He cleared his throat and said, 'I want to give you my blessing. Please always feel free to come back, but in the meantime go and spread the seeds of the wilderness. I wish you light and wisdom, and please, if we call you for help, come back to lead our trails.'

As he shook my hand, I realised this really was happening. It was a painful experience, but it was a pain I had to embrace to move on. I had made good friends at the school, and would always be grateful to my mentors – Paul Cryer, Bruce Dell, Ian Read and Baba Gumede, to name only a few.

I walked away from the Wilderness Leadership School with much sadness. But no one can stop the wind.

Part Five

———•———

FINDING A PATH THROUGH EZEMVELO

2007 to 2014

DESPITE MY SADNESS, I WAS EXCITED TO RETURN TO EZEMVELO KZN Wildlife, my first wilderness home. The organisation had a solid reputation, with a history of attracting visitors to its reserves from all corners of the earth. The so-called 'primitive' trail was one of Ezemvelo's flagship wilderness trails, initiated by Ian Player and Magqubu Ntombela, and to be a part of it is every guide's dream. I knew I had a role to play as a conservationist, and the fire to do this had burnt ever more strongly in the five years I'd spent at the Wilderness Leadership School. I had nearly completed my studies in wildlife management, and was pleased to be back in the fold of an organisation tasked with the conservation management of the iMfolozi wilderness.

I was grateful to my parents, who came to stay with me at iMfolozi for a few days to help me negotiate this change. Spending time with them after work triggered old memories of the many beautiful times I had spent here with my father. I particularly remembered the time when he'd faced down the lions with only a knobkerrie, and how much this had taught me about courage and resourcefulness. We spoke a great deal, reflecting on those experiences and on the role my mentors had played in developing me.

I was looking forward to joining my former colleagues and contributing some of the insights, skills and experience I had gained at the Wilderness Leadership School. But I soon discovered that some of the philosophical foundations that Player and Ntombela had laid had been diluted over the years. Ezemvelo's wilderness trails had become more like safari adventures: trailists came with checklists to tick off their Big Five sightings, and guides were tasked with unearthing and presenting these animals.

I could understand that, for people who had no opportunities to see wild game in their daily lives, there was a hunger to see as much as possible. And perhaps the experience of watching nature documentaries creates false expectations. But the tendency to see trails as an adventure or spectator activity did not resonate with me. I knew that most people I had walked with at the Wilderness Leadership School were hungry for the healing that nature could bring. Yet it seemed to me that coming on a trail as a spectator would block that healing power.

As I thought about what was missing in the 'spectator' approach, I began to understand that the essence of the healing power of the wilderness is not just being physically present in it. It is about finding connection with it. A person speculating for a mining opportunity, a poacher seeking a rhino, a soldier at the border will all be in nature, but they are unlikely to be healed by it because their intention disconnects them from their surroundings. Our disconnection from nature has grown in layers, beginning perhaps with the first moment we domesticated an animal to make it a tool rather than a companion. Our centuries of 'civilisation' have disconnected us increasingly from the sources and ongoingness of life, leaving us suffering a great loneliness. Encountering wild animals is a vital part of the wilderness experience. But if we encounter them as a form of entertainment, as something to be viewed or manipulated, we create a barrier that pushes us outside true connection.

Trying to understand this approach to guiding helped me to think about what was different about the kind of trails I wanted to run. I began to understand more clearly some of the essential ingredients of a meaningful and healing trail.

A wilderness area such as iMfolozi offers immersion, the opportunity to dive deep into a wild space that has no signs of human habitation. This is valuable, because signs of human habitation remind us that humans control, exploit or manipulate most of the spaces they occupy. Although the park is itself a managed space, it offers a good sense of what it is to be in a wilderness where humans neither shape the landscape nor manipulate what happens in it.

Being present in such a space is an invitation to step away from our understanding of ourselves as a dominant species and yield our power to the greater power of nature. Yielding this power disables the ego, dissolves the boundaries of self and opens us to a deep level of connection. It may be that the wilderness is created not only by the space, but also by the mindset of the one experiencing it. If we can cultivate our ability to relinquish our power and ego, our need to intervene in our surroundings – if we embrace all other living creatures and acknowledge their right to exist, even the humble ants – we could find wilderness in the corner of a neglected garden or a 'weed' growing through a crack in the tarmac.

When immersed in the wilderness, it doesn't matter whether we meet a buffalo in a mud hole or a beetle on a flower, whether the wind is blowing or the rain is falling. We connect to the animals through the scents and sounds all around us, we see their tracks, we drink from the same rivers. Our bodies know they are there, whether our eyes see them or not, and so we learn to tune in to the softer sensing of the heart. And should an animal choose to let itself be seen, we will be all the more alive to the wonder of this meeting.

To create a healing, immersive trail, I needed to step back and lightly hold the space so that people would feel safe. I needed to offer silent hours with abundant space for self-reflection and absorption; allow encounters with animals that could remind trailists of both their vulnerability and their courage; help them to disentangle themselves from the calculating brain by removing time and schedules; and give them space to connect to the feeling heart.

At the Wilderness Leadership School, I'd led groups who'd come with different intentions, but all had found healing, because we'd led the trails in a way that opened that connection: walking in silence without the constant commentary of 'bush interpretation' opened space for the trailists. Engaging people in all campsite activities, such as washing pots, cleaning the campsite, collecting water or digging for water during the dry months, grounded them in the simple rituals of sustaining life, common to all creatures. Walking without schedules helped them yield control and tune in to the deeper rhythms of their own bodies, and the slow tracking of the sun, moon and stars through the sky. All of these were part of the immersive wilderness experience that enabled, at least for a few days, some sense of the deep connection that civilisation has severed.

I resolved to lead trails at Ezemvelo according to my philosophy and beliefs, although this was sometimes challenging. Some trailists found it difficult to give up their expectations – it made them feel anxious and out of control. Some became aggressive, demanding that I produce lions or leopards for their entertainment. But I knew it would be a betrayal of my life's purpose to provide these kinds of experiences, and persisted with my approach. I was relieved when one of the groups that did annual trails with Ezemvelo reported that I was 'the missing ingredient' in the iMfolozi Wilderness Trails. This affirmed my approach, and I was fortunate that my manager at the time, Kim Gillings, was an insightful and sensitive leader who gave me the support I needed to stay with my resolve. Thanks to her, despite the challenges, I never felt unsupported. I have a deep respect for the role she played during her time in the office and her dedication to the protection of our fragile wilderness.

———

Soon after I started at Ezemvelo, I led a trail that highlighted this 'missing ingredient' and showed the difference between 'spectating' and connecting.

It was a trail I did in 2007, with a group of people from Europe. One of the trailists was a Belgian woman who was very keen to see African wild dogs – she was a scientist studying wild canines, and had done work tracking wolves in Russia and other parts of Europe.

I suggested that she *invite* the dogs to step into her inner zone.

I heard the dogs in the night, but was amazed to wake to their high-pitched squeals and yammering in the morning, as they are not often sighted. There they were, just beyond the water, still feeding on the waterbuck they had killed in the night. The adults watched us with curiosity but no fear, great circular ears framing delicately patterned faces; the pups tumbled over one another and tussled over the kill. It's rare to see these animals so close. But they had been invited, and they had come.

We watched the wild dogs leading their wild lives, the pups playing tug-of-war with the waterbuck's head. Later, a group of wildebeest came down to the river. At first they were cautious about the dogs, but when they realised the dogs were not hunting, they relaxed, and some mock-charged the pups, which chased them back. They'd sensed the wild dogs' playful energy, just as the dogs had somehow sensed the Belgian woman's invitation. We sat in the golden morning, alive with the joy that comes from a meeting with other species as equals and friends.

Inviting an animal is a subtle communication between the trailist and the animal that recognises the animal's agency and respects its wish to be seen – or not. To invite animals in this way, people need to soften their intentions, open their minds and hearts, step away from their own desires and egos.

One of the ways I helped people to do this on trails was by doing away with time. We had no schedule; people were encouraged to leave their watches and phones behind, and I never mentioned the time. This took us back to the time of our ancestors, who spent three or four hours a day gathering food and the rest of the day in storytelling and contemplation.

When you enter this eternal present, as you do when you meditate, you empty your mind and reach a state of innocence, with no stress, no intention … a soft, open space that an animal can easily step into. I remember once doing a deep solitary meditation at

As trails officer with Ezemvelo KZN Wildlife – the 'missing ingredient'

Bergplaas, a wilderness reserve in the Karoo. I sensed a presence and when I opened my eyes, I saw an aardvark snuffling the ground barely two metres away. This shy, usually nocturnal, creature had felt the soft invitation of my meditating mind and felt safe to approach me. When you reach that state of openness, you become one with everything – you become a rock, or a mountain, or water. Animals live in that state, and can easily recognise and engage with you if you are in that state too.

There is no doubt that seeing an animal that a guide has found for you will be interesting. But seeing an animal that you have invited is truly transforming.

⋅────────⋅

The kettle hissed on the fire as the smoke from the umthombothi and wild camphor wood drifted into our eyes. I shuffled forward and pushed the wood further into the fire, sending sparks flying up like a festive fireworks display. But the leaping flames lit up a face that was not celebrating. Flowing from his eyes were two streams of tears, gleaming in the firelight. I stretched my arm over his shoulder to comfort him.

Chris was a young South African who had chosen to join the Royal Air Force and serve time in Afghanistan. This trail, conducted in my second year of working for Ezemvelo, had been organised by the British military as part of a debriefing process for soldiers who had been in active combat – before they returned home to hug their wives or husbands and children, and mix with civilians. Chris needed this badly. He was suffering from having touched the blood of war, scarred by the pain and death he had seen and inflicted, tormented by vivid flashbacks of horror.

I watched him, moved by his anguish, but trusting in the wilderness to heal his raw inner wounds. His voice entwined with the sounds of the night animals going about their lives around us. Beyond the river two leopards were mating, their yowling and growling rising to a crescendo as they were struck by the sacred lightning of new life. Their act celebrated life and promised birth, in sharp contrast to the death and devastation haunting my companion.

Closer by, we could hear the phragmites reeds being shaken and uprooted by a bull elephant – we could not see him, but as our scent blew towards him he rumbled and trumpeted. Around us were the sleeping forms of the other soldiers, undisturbed by the animals or our voices as we spoke far into the night of all the troubles of the world, from nuclear weapons to the theft committed under colonialism and the ongoing oppression of African people. We agreed that the world no longer needed military heroes – if it ever had – or strongman leaders who rob their own countries and start wars that exacerbate global warming, the displacement of humans, the destruction of our forests and oceans.

Chris was devastated by his role as a cog in the military machine. But all of us are touched by human violence; most of us have ancestors who have contributed to the destruction of the earth, to the genocide and exploitation of native tribes. My Zulu ancestors were killed and taken into servitude by the British – but they themselves hunted and killed San people because the San did not recognise the concept of ownership of animals, and hunted the Zulu people's cattle as they would any wild animal.

I sipped my rooibos tea, now gone cold, and listened to Chris, struck by his grief and vulnerability. I thought of how soldiers are presented as cold and immune to emotions. How many of them are suffering this inner anguish, hidden from the world?

A crested francolin called from behind the campsite, bringing us back to our surroundings. Without our noticing, the fire had burnt down to a few glowing embers. Chris added some umthombothi logs, and blew on the embers to rekindle the flames.

'You know what, my friend?' he reflected, as the fire flickered back into life. 'The wilderness really has the magic to heal our wounds. Just being here is a therapy itself ...'

I nodded in agreement. 'Yes, I know it well. These experiences can give us courage and help us not to be ashamed of our tears – for they are tears of healing.'

The cool breeze of the coming dawn swept through the campsite, turning my thoughts to my warm sleeping bag. Chris went to wake the next watch, and I retreated for a few hours' sleep. I shuffled into my sleeping bag and gazed up at the constellations, at Scorpio rising over the hills of Ezindengeni. This name means 'hill of broken clay pots'. It was named after a battle of two mighty Zulu kings, King Shaka and Ndwandwe, also known as Zwide. Ndwandwe had declared the hill a refreshment site for his warriors, so food and beer were hidden there for them. But his warriors were defeated by Shaka's warriors. They had no time to replenish their supplies, as they were being hounded out of the area. When Shaka's warriors arrived on the hill, they destroyed all their clay pots, scattering the hillside with shards of broken pottery. I thought of all the soldiers like Chris, shattered into pieces by their experiences, like those pots – how all of us are cracked by the traumas of our lives.

I am just an ordinary wilderness guide who carries only a firearm, five rounds and some truths. The power to heal is not in my hands, but in the wilderness, in the entanglement and beauty of life. Connecting to the wilderness enabled Chris to open the door to that hidden room of his pain and release his tears – and I knew his healing had begun.

I saluted the leadership who had the foresight to arrange this trip to heal these wounded soldiers, although there was something a little cynical about patching up the soldiers' emotional wounds so that they could be released into society. I know that our own country is haunted by the 'walking wounded', traumatised by the wars of apartheid on both sides. My own father had been wounded as a soldier in Umkhonto we Sizwe, the armed wing of the African National Congress during the struggle against apartheid. He

Campfire therapy

carried the pain of those scars through his life. The outer wounds heal in time, but the inner wounds may continue eating away at the spirit for a lifetime, particularly for those white conscripts who'd been fed lies to fight for the apartheid cause that was so morally indefensible.

The walking wounded carry these wounds still, their pain spilling out so that they keep hurting themselves or our society.

If only we had enough wilderness to heal them all …

●────────●

I shuffled my back to find a more comfortable position. My hips were aching, and the cold of the rock was slicing through my sleeping mat and sleeping bag. The spotted thick-knees wailed, anxious and strident, in the darkness. Shivering, I reached into my backpack to pull out another liner for my sleeping bag. The dew was heavy on the sleeping bag, dripping icy water onto my skin as I wriggled into the liner. The fog of the graveyard shift rose from the river over the camp, thick and impenetrable, intertwining with the crackling of red ivorywood on the small fire. I lay back in my bag as the fog and fire rejuvenated me, dispelling my discomfort and resistance to the cold.

I have helped many people to conduct many rites of passage on my trails, but this trail, in the winter of 2009, was for a dear friend, Erick, who had only a few weeks left to live. The whole trail was a rite of passage, and touched me deeply.

The fog shifted and wound around us, giving us brief glimpses of the stars. I could tell by Scorpio that it was the middle of the night, as it hung above our campsite, almost in the centre of the sky. The ghostly smudges of the Magellanic Clouds hovered over the hills of Mahobosheni – the place of puff adders. I'd been told by an old friend that the hill had been given its name by the sharp bend in the river below, which had shaped the land to resemble the head of the puff adder. I could hear the staccato whooping of the African wood owl and the distant sawing of a leopard almost five kilometres away. Something heavy was sloshing through the river – a rhino drinking, perhaps, or an elephant.

I have always been acutely aware of the sounds of nature, of their power to bring peace to the ruffled soul. But Erick had made me even more awake to these sounds. I'd first met him some years before when he'd done a trail with me at the Wilderness Leadership School. We'd remained friends, and climbed mountains together. But now Erick was facing a mountain he had to climb alone. He had been diagnosed with terminal cancer, and had come to the wilderness with his family to celebrate his life with them, to celebrate the miracle of all life on earth, and to say goodbye.

I knew from my reading and conversations with Meredith Little, and Sarah from the Glencree trail, that the wilderness can offer a sacred tool to help you make peace with your journey into the unknown land beyond this life. But this was the first time I'd had the privilege of helping a dear friend go through this sacred rite of passage.

Erick spoke often of how the sounds of nature helped to release him from the stress and noisy clamour of his working life as an executive in a major engineering firm in Johannesburg. On this last trail, he was intensely alive to every sound, and I became alive to them myself to honour his experience. Not only the sounds of animals giving voice to the emotional tapestries of their lives, but also the quieter sounds ... the patter of rain on my flysheet, the whispering of grass in the wind, the crackle of the twigs underfoot. At night, the sounds were amplified by the darkness – the soft whirr of an owl's wings and the frightened quiver of its prey, the faint beating of a moon moth's wing as it brushed my cheek. When I closed my eyes on this cold night, I seemed also to sense the soundless sounds, the hard bright silence of the distant stars, the soft damp silence of the fog.

Our trail was one of cool, star-bright, moonless nights, dense with the wilderness dance of predators hunting and prey running for their lives. A rich symphony of animal calls wove around us each night, from the shrill chirping of the crickets to the deep rumbling of the lions. I lay in my sleeping bag on these cold nights and watched a dying man doing guard duty, silhouetted against the firelight, weaving the sounds of the night into a bridge to help him cross over to the other side. Finding peace in contemplation, embracing the new life that comes with dying. And so, through the deep wilderness night, he came to discover the brightness that would light his way on his last journey.

He called me a few days after the trail, but I missed his call as I was on another trail. When I returned, I played his voice message – his voice faint and distorted, saying, 'Goodbye, my friend. I see the light. I think my time has come.' And a later message from his son, saying his dad had passed away.

His family wanted to bring his ashes back into the wilderness, and I managed to organise a trail for them to do this. We walked through the wilderness, his son carrying the urn. Erick was with us – not in the urn, but returned to the great spirt of the iMfolozi. He was in the wind rushing down the hills on the moonless nights, the river swirling over our feet. It was especially in the nights that we felt his spirit, singing with the great chorus of animals, stars, trees – the chorus of everything.

On the last day, we climbed the Momfu cliff. His family cooked his favourite meal, and we ate it together. Then we stood looking over the silver snake of the iMfolozi River, undulating into the far distance through plains and hills teeming with animals. We released his ashes, and watched them blow away in the wind.

Erick often said on his last trail that he hoped to die there, to breathe his last breath in the wild beating heart of the iMfolozi. I teased him, saying, 'Please don't die now, or I'll be tied up in paperwork for the rest of my life.' But I thought often about this

So many symbolic crossings in the wilderness

wish, about what the wilderness had given him that would make him want to lay his life down in it. I was grateful that I'd had the opportunity to see the wilderness through the eyes of a dying man, to feel the clarity of his gaze, as he realised that all the busyness and compulsion of his working life was nothing compared to the beauty of the sunlight on a spiderweb beaded with dew. All that mattered to him in his last days was love: love for his family and friends, and for this vast, sprawling, infinitely varied, beautiful and miraculous family that is life on earth.

———————

When my trails manager left Ezemvelo in 2010, I was appointed as acting manager of the wilderness trails. This was a promotion – but it took me out of the wilderness, as I now had to spend much of my time in the office, dealing with budgets and orders,

managing staff problems and leave forms. We were seriously understaffed, and I soon began to find this work soul-destroying.

But it brought new opportunities, and one of these was reviving
the Wilderness Community Awareness Programme (WCAP), which
enabled non-profit trails with local community members.

The Hluhluwe–iMfolozi Park is surrounded by communities
who, for the most part, have never set foot in it, let alone walked
in its wilderness. For them, the game reserve is exclusively for
privileged white tourists. The management of the park were aware
of this problem, but were under pressure to generate income – and
trails with local communities did not do so. In fact, not only did
community trails not bring in revenue, but each trail took a guide
and backup guide out of the pool of guides needed to run the
profit-making trails.

There were some community programmes to create awareness
about poaching. But many community members saw this as a cynical attempt to stop them from poaching, rather than a real effort to
address the historical dispossession of their land and their ongoing
exclusion. These programmes did not dispel the animosity local
people felt towards the park, especially when lions or hyenas occasionally broke out and killed the livestock that these people relied
on for their survival. Compensation was paid, but it was often de-
layed. Because of this alienation and the extreme poverty in the
area, when poaching syndicates came with big money they did not
struggle to recruit poachers to shoot rhinos – although many community members were opposed to this, and helped the park by
giving information about poaching operations.

I was determined to strengthen the community's access to and
benefits from the park – not just because the community might
resort to poaching, but because I believed it was their birthright to
walk through the wilderness where their ancestors had walked, to
experience the richness of wildlife and nurture their souls with
solitude and silence. My chance came when I was appointed as

acting trails manager. On sorting through the office files, I discovered that the WCAP had the budget to run at least one trail with locals each month.

I jumped at the opportunity to revive this project. Here at last was a chance to open the gates of the park to young people in the surrounding communities, and to break the long, high fence of exclusion that had deprived our locals from the benefits of the park for so long.

I went into the community, and spoke to parents and teachers. To select learners for the trail, I asked them to write an essay about rhino poaching or climate change. I chose the best one, the worst one, and a few in between. I wanted a range of learners – the wilderness is not for high achievers alone. Someone who is failing may need it even more, for being in the wilderness can help them find their strengths. The only sadness was that I had to choose eight participants – I longed to take them all. In the winter of 2010, I finally realised my dream, and set off on a wilderness trail with eight Grade 11 learners from my community.

When I gave my safety briefing at the start of the trail, I could see both fear and excitement in their eyes at the thought of sharing a path with lions and elephants and sleeping under the stars. As we prepared to cross the iMfolozi River, I gave my speech about how we were crossing into the wilderness. But I knew we were crossing so much more than a river. We were taking a small step into a new way of being in the world, a world where youth were empowered to work towards a society no longer blighted by racism, sexism, inequality and dispossession.

We set off on a narrow path through leafless trees and naked bush. The Cape turtle dove seemed to say, 'Work harder, work harder.' The wind whispered songs to us through the wild olive trees, sending showers of leaves falling to the ground to crackle underfoot.

We passed fresh hyena scat, hung with buzzing glossy green

flies, alongside older droppings that had turned white. I pointed this out to the young teenagers, explaining how the fresh droppings fed the flies and nourished the soil, how the scat turned white because it was rich in calcium from the bones the hyena had crunched, how these droppings were also a source of calcium for tortoises and land snails, who needed it for their shells.

'This is a very useful animal,' I said.

The youngsters were astonished that a hyena could be seen as a useful animal, that even its dung was useful. In their villages, hyenas were seen only as brutal creatures that break the fences of the reserve to steal goats and calves. Now, at last, I had the opportunity to help break other fences, the fences of seeing wild animals only from the narrow perspective of the good or ill they bring to humans. To open people's eyes and hearts to recognise nature as a great teacher, and wild animals as our kindred spirits.

In the deep quiet of midday, as we were resting by the river, a herd of elephants came through, disrupting the silence and flooding the river with their energy. The calves gambolled in front of their mothers, lurching on unsteady legs. Young bulls ripped up the sweet green grass voraciously, like teenage boys tucking into food after a soccer match. The mothers drank calmly, their strong gentle presence holding the boisterous young ones in check. The young trailists watched the scene eagerly, laughing at the tussling of the young bulls that mirrored their own games of wrestling and stick-fighting.

'I wish we could stay here forever,' one of the young women said wistfully.

I wished they could too, but I hoped that they would carry some piece of this in their hearts forever. We spoke softly about the elephants, about how they look after one another in the herd, about what they can teach us as humans, about recognising our kinship with these animals. How cruel it was that our history had kept these young people from knowing this wilderness – but here at least they could briefly experience that powerful, grounding sense of belonging and connectedness that has saved me so many times.

The hot northerly Zululand wind carried our scent to the elephants. The herd moved restlessly, alert and wary. The mothers stopped drinking; the calves stopped playing and ran between their mothers' legs. A lonely waterbuck bull watched us with sharp intensity; a warthog family erupted from the reeds and a flock of weavers flew up to the other side of the riverbank. We moved away, to allow the animals to regain their peacefulness. The youngsters walked in silence, as they had been told, but I knew they were all thinking deeply about what they had seen, and would have much to say about it that night.

This trail, like the many that followed, was a great success. Many of the youths who came on these trails developed a passion for conservation and went on to build careers in it. All who attended were given a rare opportunity to experience the simple healing power of connecting to the wilderness, and to extend their intellectual and emotional horizons. Our nights were filled with deep discussions under the stars, about how we could build a place of unity and equality. But during the days the trailists were simply joyful young beings in a wild place, touched by the sacred soul of the wilderness, alive and free.

———•———

The trails with local youths gave them an opportunity to connect not only with plants and animals, but also with their own histories, as I discovered in the winter of 2010 when on trail with a group from uCilwane village, just south of the park.

A dry winter haze hung over the confluence of the Black and White iMfolozi rivers. It was a Sunday, and the sounds of drumming and people singing in a faraway church followed us. They invoked the ancient battle cries of long-dead warriors, slowly fading as we pushed deeper into the wilderness. We followed a well-worn path, zigzagging through a forest of wild camphor, the smell of fresh buffalo dung heavy in the still air. As we approached the confluence we came to Shaka's hunting pits, dongas that Shaka's warriors dug more than two hundred years ago. This was the site of the great annual royal hunts, where warriors chased animals into the pits. The hunts were only carried out in winter, as in summer the animals were calving. The animals the warriors killed provided meat for the community, and hides for the warriors' battle dress and other clothing.

Beyond the pits we could see the black slate shelves of the Black iMfolozi, fringed with tall phragmites reeds. All was still. Even the birds were silent; not a leaf trembled. There was a strong presence of lives gone by, and I felt goosebumps rising.

One of the young men on the trail began calling, speaking in a trancelike state with the voice of one who is channelling the ama-dilingozi, the ancestral voices. Another asked for permission to recite the izibongo (praise poems) for Shaka. When I agreed, he began singing the king's praises:

UDlungwane kaNdaba UDlungwane lombelebele Oluhlunge emanxulumeni kwaze kwasa amanxuluma esebikelana Ilembe eleq' amanye amalembe ngokukhalipha UNodum'ehlezi kaMenzi uSishak' asishayeki – Umta Ka Senzangakhona, uShaka ngiyesaba nokuthi unguShaka ngoba uShaka kwakuyinkosi yasemashobeni Bath' uShaka kayikubusa akaykuba Nkosi Kanti kulaph' ezakunethezeka khona. Umlilo wothathe kaMjokwane Umlilo wothathe ubuhanguhangu Oshis' zikhova eziseDlebe Kwaye kwasha neziseMabedlana.

Sitting with these millennials – born in the infancy of the twenty-first century – on this site of rich and tangible history stirred my heart. The sons and daughters of my community were touching the souls of their ancestors. I heard a young woman say, 'This place is giving so much to me. I feel rich but I have nothing.'

These youths did indeed come from poor homesteads. They were ending their school careers with few prospects. And yet this young woman had experienced the inner wealth of a soul blessed with a deep knowledge of who she was and where she had come from. By walking the paths of their ancestors, these young people could feel the presence of those who had lived in a time before colonialism, before their lands had been taken, before their identity and self-worth had been trampled by English missionaries, before the colonial system had created poverty to force people into the labour market.

This inner richness gives us the generosity of spirit to share our food and water with those with a greater hunger and thirst. It gives us the wisdom to value not the soaring mountains but the quiet valleys, for they are the foundation of the mountains. To value not the spectacular waterfalls but the humble underground streams that bring water and life to arid plains. It teaches us to embrace all, from the trumpeting elephant to the hardworking ant, to find the sacredness in the smaller life forms. It is our inner compass, showing us the direction and purpose of our lives.

It's intangible, perhaps something we can only know by its absence – the despair and grief and self-hatred that pours into its place when it is gone. And it was fast diminishing in our communities, worn down by poverty and the cruelty of a society that worships material wealth while denying most people the means of acquiring it.

As the shadows lengthened we gathered our packs and left that place, watched by a lone nyala bull. My heart was full of the miracle of the day. I had seen how deeply these youths had been touched by the wilderness, and by the powerful ancestral spirits slumbering in the rocks and gnarled roots of the trees. I knew their lives were difficult, and would still be difficult, but their hearts had been infused with a light that would empower them to dance through their difficulties. I felt confident that they would carry this light back home with them.

The healing power of connecting with the wilderness is needed by all, regardless of race, but perhaps needed most by those who have been most wounded by history. As we made our way to the campsite for the night I knew that this was truly my life's work.

———————•———————

When I took up the position of acting trails manager, I came to realise that many of the guides were demotivated. They were suffering under the 'walking safari' model of nature guiding, which I had noticed was creeping into the wilderness trails.

I spoke earlier about how this 'walking safari' approach diminishes trailists' experiences, blocking them from being fully open to the animals they encounter, and about how it can be invasive to the animals. But it is also toxic for the guides. It reduces guiding to a shallow experience, like dropping a plate of food before a diner at a restaurant.

Luxury lodges advertise themselves as guaranteeing Big Five sightings, enticing guests to pay exorbitant fees with promises of seeing leopard or lion. The guide becomes a machine, trying to summon up the Big Five from twenty-five thousand hectares of wilderness. Those who do not 'perform' may lose out on tips, and may even lose their jobs.

This is unrewarding work, soul-destroying for guides who come into conservation with a passion for nature. Many get discouraged, and leave conservation if they have other options. I have been doing wilderness trails for twenty-five years and have never got bored. In fact, each trail refreshes me, and brings me to renewed life.

From my first day as trails manager, I tried to bring in a different culture. I told the guides, 'I am not here as a king. I am not here to judge you about how you conduct your trails. I am here to learn from you. If you can learn from me too, we can build a strong family, which is not divided by who is superior and who is not superior. I am here as part of you, as part of a family who wants to do something so that we can help our world.'

The guides asked what I meant – they had no understanding of how taking people into the wilderness is healing for the world. I explained that taking people into the wilderness is deeply healing, that guides can be like therapists who facilitate an experience that will help to restore people's mental health, and may even save someone who might otherwise have committed suicide; that guides can also help to heal the relationship between humans and nature, a relationship that has broken down so badly that our future on earth is in danger.

I needed to teach more guides, and grow the family that would enable 'soul therapy' on the wilderness trails. I began this work while I was acting trails manager, but I continued mentoring guides throughout my time at Ezemvelo as I could see what it brought to them.

To do soul therapy work, the guides needed to experience the healing power of the wilderness themselves. I focused on building their awareness, but once their eyes were open to it, nature became their teacher. This was the case with Siphesihle Ncgobo, a young guide from Pietermaritzburg.

While on trail as my backup guide in the summer of 2010, Siphesihle had a powerful encounter with an animal that transformed his way of seeing things.

The heat had been building up for days. The acrid air singed our nostrils; the white-hot sky pushed down on us as we walked over ash-dry grass, our boots thick with dust. It was still scorching when we reached the camp, but I could see dark clouds rolling in over the hills in the distance, and knew that at last we would get rain later that night.

I decided to put up the flysheets while we still had daylight, rather than battling with them in the bucketing rain in the dark. I asked Siphesihle to take some of the trailists to gather firewood. When they were some way off, I could hear from their agitated voices that they had found something, but I was too busy to leave my flysheets and investigate.

When they returned some time later, Siphesihle's wrist was bleeding from a bad gash – but there was a lightness in his step and pride in his eyes. He explained that they had come across a white-backed vulture in a tree. The creature had a rope around its neck – it seemed to have escaped from a trap, and had managed to fly into the iMfolozi. But it was exhausted, perhaps dehydrated, and seemed unable to breathe well.

Siphesihle immediately climbed the tree to help it. The bird panicked and fought him off, but he held on to the rope until the bird's struggles began to weaken. Slowly and painstakingly he managed to untie the knot, all the while being pecked and lacerated by the bird's powerful beak. He continued to work, with blood dripping from his wound, until the bird was free.

On the day before, we had seen a group of these magnificent birds splashing in the river to cool off, fluffing and twirling their feathers in a spray of silver droplets. Siphesihle had told the trailists about them, about how they are under threat from farmers and from traditional healers who use their eyes for muthi (traditional medicine), believing that it gives people the power to see into the future and pick winning lottery numbers. Siphesihle spoke passionately about these birds. It seemed his passion had called this creature to him, as if the vulture knew that here was someone who would surely help it.

Siphesihle was one of many young guides I was mentoring to conduct immersive wilderness trails. Because I wanted them to have their own transformative experiences, I made a point of inviting whichever backup guide was on a trail with me to listen in on conversations, to be part of the solitude and meditation exercises. When they saw how the trailists paid attention to what I was saying, or wrote down quotes to display in their homes to keep them inspired, or said things like, 'I would like to bring my son to hear you speak like this,' the guides started to understand that the work I was doing was different. And some of them began working hard to develop the understanding they needed to lead trails in this way – guides such as Siphesihle, Phumlani Mthembu, Sabelo Mdlalose, Bheki Mathe and Magnificent Ncube.

It was difficult at first, as they had come from being field rangers. It was a big adjustment to move from law enforcement to working with trailists. Some did not even have a matric certificate, and lacked the confidence to interact with white people in a language that was not their mother tongue. But I encouraged them to participate in activities like cooking, which created opportunities to interact with the trailists.

I also took the time to talk with them about global environmental issues, such as climate change, habitat loss and mass extinctions, and challenges that plague our society, such as poverty and social inequality. This awareness of socioeconomic problems helped them to understand what causes the diseases of the soul from which so many people are suffering, diseases like loneliness and alienation. And it helped them to understand how walking in silence through an umthombothi forest, while listening to the calls of doves and bushshrikes, can offer people a sacred space to heal these diseases and come to new life.

I helped to widen their horizons by encouraging them to start reading books that would enhance their knowledge about guiding and philosophy, and give them the confidence to participate in these discussions themselves. I borrowed books from the Wilderness Leadership School and lent out my own books. My copy of The Holy Man soon became dog-eared and battered from passing through so many hands.

At this time, the Bergplaas Nature Reserve, near Graaff-Reinet in the Eastern Cape province, was offering the Spirit of the Wild programme to guides employed by Ezemvelo and SANParks, the organisation managing South Africa's national parks. Princess Irene van Lippe-Biesterfeld of the Netherlands, who was disturbed by the shallow consumerist approach to nature safaris that she had experienced when visiting reserves and game lodges in South Africa, started this initiative. She bought a farm in the Eastern Cape for training guides and others into a more intuitive way of guiding.

I had been fortunate enough to attend a Spirit of the Wild course while I was at the Wilderness Leadership School, and knew it would be an invaluable experience for the guides, some of whom had never even left KwaZulu-Natal. The Bergplaas programme subsidised the guides' transport and training, but Ezemvelo needed to release them for the three non-consecutive weeks required for

With Bonangiphiwe Mbanjwa

the programme. We sent a few guides on this programme.

I also encouraged the few women guides, and mentored Bonangiphiwe (Bona) Mbanjwa and Mphile Mthethwa. Guiding has historically been a male-dominated career, with women reluctant to come forward due to gender socialisation. Bona grabbed the opportunity, and started reading the books I lent her. She told me that after her father had been killed by gunmen, the only thing that had kept her alive was seeing the beautiful flowers of the sickle bush growing at her homestead. When I went with her to Richards Bay to collect her certificate she bought herself her first book.

The guides began to flourish because their work was so much more sustaining for them. It fed their souls, which enabled them to feed those of others.

I remember conducting a meditation on the last day of a trail with a group of Australian lawyers. I was reflecting on all the things we had encountered, and highlighted Phumlani's work as a backup guide, collecting firewood, cleaning the sleeping bags and so on. By the end of the meditation, many people were in tears, including Phumlani. When I asked him later why he'd been crying, he said he'd been moved by my acknowledging his contribution to the trail. 'I never thought leadership would recognise me. Thank you for acknowledging me and giving me this gift,' he said.

The gash that the white-backed vulture gave Siphesihle healed over, but it left a scar. I told him that this was a mark of great courage, a tattoo of love and passion for protecting these iconic birds – that it would always remind him of his passionate concern for the creatures that share our world. I found it so moving that someone who'd grown up in the litter-strewn township streets had so much passion for nature and wild creatures that he'd been willing to suffer injury to save one.

Later, Siphesihle wrote a poem about vultures flying over the world, seeking out wisdom with their far-seeing eyes. This was my hope for the guides I mentored – that I could help to free them from the ropes that 'walking safarism' had placed around their necks, and strengthen their wings and widen their gaze, so that they too would ride the high winds of their lives and gain healing wisdom wherever they flew.

———————

In the winter of 2011, I was walking with a group from Cape Town towards a thicket near a stream. The pale, dry grass crackled under our feet; a white-backed vulture hissed and cawed as it flew up from a torchwood tree, shedding a few feathers. I had chosen to have an 'empty day' – a day with no theme or agenda, devoted simply to receiving whatever the wilderness had to offer. My mind was a wide, empty plain, without a single thought; my soul was light, glowing like the silver gleam of sun on the river that had greeted me that morning.

I knew that this openness was good for inviting animals. I would soon discover, not for the first time, how such openness can also save lives.

We were approaching the Mphafa Stream, which we needed to cross to get to our campsite. The thick bush around the stream is notorious for harbouring the most dangerous animals, from black rhino to black mamba. As we drew near, the bush closed in on us, until we were walking through a tunnel lined by a dense wall of creepers, with shrubs and reeds looming over our heads so that sometimes we had to bend double to get through. A multitude of small, straggly paths led off into the reeds. There was fresh lion spoor on the path – a pride of about ten with cubs. We had heard them roaring from this forest in the night. The stillness of the bush was ominous, as if it were holding its breath, waiting for two objects to collide. A crested guineafowl flew up abruptly from the bush, startling my group and causing one of the women to stumble. A group of vervet monkeys looked down on us from the high branches of a sausage tree.

When the path opened up just before the stream, I felt relief to be free of this claustrophobic tunnel. But then I felt something else: a tingling deep inside my chest, which spread like an inflating balloon pressing against my lungs. My shoulders stiffened, and goosebumps rose on my skin. I indicated to my backup guide to walk the trailists back under cover.

He'd scarcely done this when a black rhino and calf bolted from the bushes, charging straight towards us with a great cracking of vegetation and loud snorts like a steam engine. All broke into chaos – some tried to run behind the bushes, others to scramble up into the low thorn trees. One woman yelled, 'Let's go home!' referring to the campsite where we had left our equipment that morning – which, even though it was deep in the wilderness, now felt like a safe haven to her. I shouted to head off the rhino – it changed course and charged towards the stream. But as it crossed, it disturbed the pride of lions that were lying hidden from our view in the dry bed of the stream. The lions leapt up and careened towards us, growling and snarling, the cubs hissing in alarm like domestic cats. Above it all, the monkeys shrieked from the sausage tree.

Somehow we kept the group together, until the lions took in the situation and retreated. When we had composed ourselves we continued on a different path, and crossed the stream further up. As our fear subsided, we began to look around us, filled with gratitude for the beauty of the day. We had been plunged into terror, made acutely and painfully aware of our vulnerability, but we had survived. This confrontation with the edge of life had given us new eyes to take in the boundless blue of the sky, the high circling of a vulture, the grass rippling in the wind ... we were alive, with a renewed intensity, to the limitless wonder of the world around us.

As we walked on, I thought of the powerful warning I'd had of the rhino's presence, and gave thanks that my body and heart had been open enough to receive it – without this, we would have walked straight into the animal. A startled animal is the most dangerous. It is unpredictable, obeying only its adrenaline-charged impulse to fight or flee. And the more dangerous the animal, the more inclined it will be to fight.

But the rhino was not the only danger my instincts were warning me about. Had we continued on the path and stumbled on the lions concealed in the stream bed, we'd have faced a far greater danger, particularly as there were cubs present – which the lionesses would have protected with everything they had. I wondered if the rhino wasn't itself warning us about the lions.

Animals have a strong 'sixth sense', which they rely on to keep themselves safe. But our human senses have been dulled by overstimulation, and reliance on electronic devices. Now, if we are in trouble and need someone, we just phone, where once we might have sent a telepathic message for that person to come to us. We use Google Maps to find our way, instead of relying on our sense of direction. And perhaps, because we live behind our high walls, insulated from the natural world, no longer alive to either its wonders or its terrors, we have allowed this sixth sense of our own to wither from disuse – although you might still feel it when you walk down a deserted street, see someone approaching you and 'use your gut' to decide whether they are a threat. Many of us have had the experience of thinking about a loved one, then getting a call from them or hearing that they have passed away. But the modern love of reason makes us reluctant to trust our intuition, and we dismiss these experiences as odd coincidences.

This perilous situation was a humbling experience, reminding us yet again that animals are not circus performers for our entertainment but complex, beautiful creatures to be treated with tremendous respect. As much as we like to imagine ourselves at the top of some species pyramid, we too can be prey – and those few moments of terror gave us a window into the mind of a prey animal fleeing for its life. These reminders help us develop the empathy we need to repair the cracks between our species and the rest of the natural world.

Later that afternoon, bathing in the river at our campsite and contemplating the day's experiences, we saw the same pride of lions come down to drink about eighty metres downstream. They played in the water, sharing the river with us, as relaxed with us as we were with them. We lay in the water under the warm afternoon sun, contemplating the beautiful complexity of this gentle encounter with the same animals we'd fled from in terror that morning.

———

The black rhino was a sharp reminder to be open to my intuition, and to listen to my inner voice. It came at a good time: my inner voice was warning me that working at Ezemvelo was not the way to pursue my life's vision.

After eighteen months as acting trails manager I had decided not to apply for the permanent position, even though I'd been told I had a good chance of getting it. There were many reasons for my reluctance, but most importantly I was disheartened by the endless bureaucracy that the job involved – the plague of government institutions. My feet would start itching after long hours at my computer, and my heart would tug me outside to wander old animal tracks under a wide sky. But I was buried in an avalanche of reports and budgets and forms in triplicate.

I welcomed the new trails manager when she was appointed, and also welcomed the opportunity this had given me to think hard about whether I wanted to make a career in management. Not only was I not convinced that I wanted to spend my days managing others, but I was also concerned about the pressure put on park management to use the wilderness as a 'resource' for generating profit. This is a weakness of the conservation model in many parts of Africa, and means that many parks are understaffed and struggle for the resources they need. I knew that there were many who worked for Ezemvelo who did not share this view and struggled with balancing the pressure to be profitable and the conservation needs of the park, but I did not want to be forced to make decisions driven by profit.

My vision for my life was to bring healing to the people on my trails and to my community by sharing the wealth generated by the trails, and to reknit the fractured connections between humans and nature. Ezemvelo had brought me many opportunities, and I had worked with many passionate and inspiring people. But my heart was telling me that I should no longer be part of this organisation. I needed to put my faith in Great Mother Earth and start my own organisation that would focus on healing, immersive wilderness experiences and giving back to my community.

The warning, the tingling, the pushing and pulling were growing in my heart, but it would take two years before I was ready to walk away. One of the people who helped me to find the courage to pursue my vision was my old friend, Richard Knight.

———————

It was nearing the end of a dry winter, in the following year. As we set off on the trail, I could already smell the earth's eager expectation of rain; the wind was coming up, sending dark clouds scudding across the sky. The liquid call of ufukwe, the Burchell's coucal, in the reeds warned us that we would soon be soaking. We could see the approaching showers blowing across the far hillside, and by the afternoon they were upon us. Our nostrils filled with the rich scent of the dust welcoming the raindrops, the wind and clouds teamed up to open the skies above our heads, patches of fog drifted across the landscape. The dry riverbeds filled with trickles, then streams, then gushing torrents of water. We were soon facing the discomfort of water dripping down our necks. Our boots and legs were drenched, and our wet clothes tugged at our limbs. But the land was dancing with the joy of the rain, the grass nodding and swaying in gratitude, the air filled with the sound of crickets and frogs singing with jubilation.

I was walking with a group from Menzies Aviation, a UK-based company. Richard had organised the trail. He had kept in touch with me after the trail we'd done with his son Tegan while I was still at the Wilderness Leadership School. He'd continued to support me by sponsoring my training at the school and my wildlife management studies, and now he was helping me again.

Richard ran a management consultancy called Maasai Camel, which works with clients from all over Europe. The consultancy had been approached by Menzies Aviation to do a team-building exercise in Africa. The company was growing rapidly and wanted to do the trail to help build cohesion among the team members. Richard asked if I could offer something that included a wilderness experience and some community work. I suggested that he arrange a wilderness trail through Ezemvelo, which I would lead, and I would then, on my own account, arrange a community project, as Ezemvelo did not offer these.

We discussed the value the trail might bring to the team. I suggested that it could help rekindle the team's human spirit, which may have been buried under the pressure of their work; that it could revitalise them, and reawaken their intuition. This soft skill of thinking with the heart as well as the brain was not valued in our society, but it could be a powerful intelligence in the corporate world, as well as in all other aspects of life.

I had seldom been on such a wet trail. The rain would not stop. We fell asleep and woke up to the patter of drops on the flysheet; we walked through sodden grass under dripping trees, mud underfoot, a heavy grey sky above. There was miserable discomfort in the group – our clothes were wet, our sleeping bags were wet. I had to wake up and fight with the fire each night to keep it going. I wondered whether the group would give up and ask to return to the lodge for the remainder of the time, but their spirits were high, and they were willing to embrace the rain and keep going.

We saw some animals, despite the rain. A giraffe followed us from the first day, always visible, its tall, dappled form appearing and disappearing through the grey mist. It stayed with us until the end of the trail, as if it had appointed itself our guardian.

We came across a pride of lions – they were drinking at the river, and retreated into the reeds as we approached. I could sense that the lions were not aggressive. As we walked past, they emerged from the reeds, and watched us calmly.

Later, we used this experience to talk about intuitive management. I asked the group if they'd been comfortable with the lions. Some said they hadn't, but had trusted my judgement. They asked me how I knew the lions were not a threat, and I explained that I could feel in my heart that the lions were inviting us to walk past. This led to a discussion about how you can use your intuition to make decisions, when you should trust others to make decisions for your company even if they seem like they may be harmful, and how you can sharpen your own sense of intuition by being in contact with your feelings and bodily sensations.

The group kept going through the rain for the entire four days of the trail. When we discussed it afterwards, many of them said that they were not enjoying sleeping in the cold, but they wanted to keep going because their colleagues had been so looking forward to the trip, dreaming and talking about it for months. It seemed that most of them were sticking it out for the sake of their colleagues. This led to a good discussion about the power that comes to a group when people are willing to put up with discomfort to support others. They had managed to make friends with a bad situation, recognising that even this would bring an opportunity to learn and develop.

I tried by all means to dispel any notions that we were 'conquering the weather'. I explained that we were not in the wilderness to prove that we were 'stronger than the weather'. We were there to be a part of the wilderness, to walk through the rain as the lions and elephants and giraffes were walking through the rain. We were there to share the collective breath of the living wilderness around us.

The day after we finished the trail, the sun came out. The earth was beautiful everywhere, the colours bright; everything seemed newly washed and sparkling. As we drove through the park, we saw the impalas pronking and running with the joy of it. The air was alive with birdsong welcoming the sunshine.

We drove out of the park, and turned off the R618 onto the long, winding dirt road that would lead to my village in the district of Hlabisa. The Kombis pulled up outside my homestead, between the houses and the cattle kraal. The group climbed out, and looked around them, taking in their first impressions of an indigenous village in South Africa.

The Menzies Aviation group in our village

This was the second part of the team-building. They would spend three days and two nights contributing to projects to uplift the community. In this part, I was no longer Sicelo from Ezemvelo – I was there as a community facilitator. We had discussed the project in community meetings in previous months to decide which interventions could bring the most benefit. We had settled on a preschool, and water troughs for the cattle and other livestock. Menzies Aviation had sent funds ahead for us to start work on the preschool, so that by the time the team came, the building would be in place. But they would do the painting, and put up the play structures and fence.

We created a work team for each project. The people from Menzies were teamed up with members of our local community who had skills and experience in bricklaying, plumbing and painting. We told the group to work out what had to be done, and to select leaders for each aspect of the project. We told them to work to their strengths: if you're a good organiser, make a list of tasks and order supplies; if you're good at plumbing, identify which

pipes and connectors you need; if you're good at painting, visualise how the preschool is going to look and list the paint colours you need. We were a little worried that the Menzies people might take over the leadership, but they recognised that many of the community members had better practical skills and were happy for them to take the lead.

The preschool, the fruit of the Menzies group community project

For three days they went at it, wielding pickaxes, mixing cement, hammering nails, laying pipes, plastering ... In the evenings, they would help my wife and mother prepare food and clear up after the meal. As I watched them work, I could sense the animosity between the local people and the park begin to melt away. Tension was high at the time, as lions had broken out two weeks before and killed some cows. But many villagers came up to me and said how good it was to see visitors to the park coming to do something positive for the community, to see them living the way we lived, using the long drop, sharing meals, walking on our ancestral paths ...

Sometimes the simplest gestures can begin the process of healing the wounds of the past. My father was very conscious of how our people had been oppressed, of how resources in Africa had been stripped and stolen by people from European countries. But he told me he was pleased to see this effort being made to give something back.

At the end of the project, my father gave the Menzies delegation a goat. They slaughtered and skinned it, with the help of my brother, and cooked goat stew. We invited all the villagers who had worked on the project to join us. By then, people were relaxed with one another, after having worked together for three days, and could laugh together and share jokes and stories, while also congratulating themselves on how much they had achieved. There was singing and dancing, and we shared some traditional beer that my mother had brewed.

My father gave thanks to our ancestral spirits, and to our visitors. He told them that he considered them the shining moon, for they had come at a time when the village was facing darkness.

As I sat watching this, my heart was full. This is how a global village is built, I thought. With goodwill and selflessness, with open hearts and hands. We needed to bring the 'abundance of rain' from the park into the community, to give the community a small stake in some of the profits generated by the park. I realised again that I could not achieve my true vision while I was working for Ezemvelo. As I watched the sun going down over the hills across from my homestead, that same sunset that had once made me weep as a child, I knew that the sun was going down on my time at Ezemvelo too.

———

Every day spent in the wilderness holds deep mystery and beauty. But some days take you through a doorway that will change you forever. Scientists may doubt the story that I am about to tell. But this is what happened, and for me it tells us that there are energies in our world that are too subtle to be measured by our instruments.

It was a mild day in the middle of the winter of 2012 – mild for me, but I could see some of the overseas visitors were feeling the heat: they were flushed and stopped often to apply sunscreen. We were walking through an open meadow of red-gold turpentine grass beneath a wide, cloudless sky. This heavily tufted grass is rich in oils, which makes it unpalatable to animals. But it is a very useful grass as it grows in disturbed ground and stabilises the soil, preventing erosion.

It will often grow on the site of a former homestead, so the grass in this meadow was a sign that this area was once a home to humans. There were other signs too – I'd previously seen pieces of broken pots, and an old broken grinding stone – but I did not mention this to my companions while we were walking through there.

On this day, the only one present apart from us was a lone wildebeest bull, who was jumping about and sneezing, uttering short gruff barks as if warning us to keep away. Some of the trailists following me began sneezing too – many people get hayfever from this grass. Brian, a trailist from Australia, was sneezing, and breathing heavily.

Then he began to wail.

It was a wail of unrestrained and inconsolable grief. It was the wail of a mother who has lost a baby, or a baby who has lost a mother. His voice was not that of a man, but of a young child. Tears poured down his cheeks, his body shook with sobs, and his wails echoed through the hills.

Everyone stopped, looking confused and shocked. I thought perhaps he'd been overcome by some memory or trouble in his life: I knew that being in the wilderness can open the doors to rooms we have kept locked. I stopped the trail, and sent everyone into solitary meditation – this involves sending people to sit in a spot where they cannot see the others, but I and my backup guide can see them all and make sure that they are safe.

When Brian had calmed down, I called everyone together for an indaba under an old Kei apple tree, and invited them to share whatever was on their minds. Brian told us that as he'd been walking, he'd heard an infant wailing loudly with desperate sorrow and grief, as if she'd been abandoned or was being hurt. The child's distress was so overwhelming that he'd broken into wails and sobs himself, and could not contain them.

He asked me if people had ever lived in that part of the iMfolozi, and I told him that artefacts and a broken grinding stone had been found there. He said that perhaps the spirit of an infant had been trapped there. It felt to him that it was a girl baby who had died during the birth process. He asked if we could collectively do a ritual to help this distressed spirit. I agreed, and asked my backup guide to walk around where we were sitting and keep watch, as this area was on a route used by elephants.

We sat in a circle, and Brian led a visualisation. He invited us to picture a colour, then he said, as if speaking to the spirit, 'I am here to help you and communicate with you. Perhaps a divine spirit sent me here to rescue you from this sadness. Use me as a bridge, follow any light, choose any colour you see here, feel it as a linen to comfort your skin, or as a glass of water to quench your thirst, or feel it as a bowl of fresh food to satisfy your hunger. Follow that light or colour until you see a gateway. You will see someone holding their hand out to you. It might be your parent or grandparent – follow that person … whoever you see will take you to your family.'

He asked for herbs to burn. The iMfolozi does not have imphepho, but we found wild curry and wild olive. We burnt these, and he invited the spirit to step through the smoke to cleanse itself, and follow the light beyond. Then we left that place to the wildebeest, who was still there. As we walked away, the animal was rubbing the gland beneath his eye against a tree to mark it with his scent.

On the following day, Brian asked if we could walk through that area again. We went there, but he said he could not feel the distressed baby. He said it felt as if the spirit had found some peace. While we were there, we found traces of old graves, which were traditionally marked by piles of small boulders. We could see that they were graves, although some of the boulders had been moved and scattered by baboons looking for insects.

After the trail, I went to visit an old man called Mkhuleumeleni Sangweni, to ask whether he knew anything about the homestead that had been there. He said he remembered, from when he was a young boy, that a family had come to settle in his district from that place. He said there were four wives in the homestead, and he believed one had given birth to a baby who had died in the birth process. He said that perhaps when the people had left that homestead, they had not performed the rituals to take the baby's spirit with them – or perhaps other rituals had not been performed. In Zulu culture, ceremonies are conducted for a baby who dies at birth or is miscarried, as you would do for a living baby. You do the imbeleko ceremony, named after the imbeleko, the animal skin that mothers used to carry their children on her back. In this ritual, a goat is killed and a bracelet made from the goat's skin is tied around the child's wrist, to keep it from harm. Later, when the child would have been about fifteen, you perform the ukuthomba ceremony for them, and so on. Baba Sangweni was suggesting that these rituals may not have been performed.

I was deeply moved by this experience, for many reasons. The emotion that Brian had expressed when he'd cried so piteously spoke of such terrible grief and distress, and that in itself was very moving. But it also moved me because I thought of my own brother who'd been lost after my mother had suffered a miscarriage. How grateful I was for the buffalo thorn, which had enabled us to collect his spirit and bring it home.

Brian was involved in fighting for Aboriginal rights, defending their sacred spaces and helping to revive their cultural practices. Although he was connected to nature and the spiritual world, he told me that he had never experienced anything like this before. We spoke into the night, reflecting on the power that comes when you connect with the wilderness.

A few days after this, Brian and I were driving through the reserve when he asked me to stop the car.

'I can feel something here,' he said. 'It feels like there were souls here who were neglected when their family moved away.'

I was astonished when he said this, as the place he'd stopped me was called Bhengu's homestead. It was a homestead whose people had been displaced when the reserve had been extended. People had been given one month to pack up and move, so there was no time to do the rituals for their ancestors who were buried there.

This experience showed me again that being a true wilderness guide is a calling of the soul, even if there were some at Ezemvelo who saw guiding as an unskilled job, something you use as a stepping stone to 'bigger things'. Wilderness guiding *is* a bigger thing, a powerful thing, connecting you not with the power of status and money but with the ancient, much more fundamental and redeeming power of life.

I knew I had developed ways to facilitate trails that created the openness that people needed to strengthen many connections: to their inner lives, to wild animals and plants, and to the divine and the spirit world. I believe it was because this trailist, Brian, was in a state of this openness that he could receive the messages of this infant spirit and, later, the lost spirits of the Bhengu clan – that my own openness had led me to decide to take the trailists on that route on that day, so that together we'd been able to bring the child comfort and lead it out of the darkness.

Much as that infant spirit connected me to the sadness of my own brother, it brought me a new lightness of being. For it helped me realise that I needed to stop wanting acknowledgement of the worth of my work from those in management who could not see it. Such acknowledgement might feed my ego – but the work I was doing fed my soul. I needed to place my faith in my own experience and that of the trailists. It was time for me to leave the darkness of my doubts.

A few months after this trail, I had another experience that helped me to find my path.

⎯⎯⎯•⎯⎯⎯

In the summer of 2013, I was walking with a group of guides from Germany, Switzerland and Austria, who had come to South Africa for a wilderness guides conference and decided to experience some Zululand wilderness for themselves. They were hungry for a nature experience, as most lived in cities.

Dark clouds were roiling above us as we crossed the waterless White iMfolozi River. The river sand carried the scars of a long, dry winter – tracks left by the hooves and pads, paws and claws of thirsty animals, and the sculpted scrapes and holes from animals digging in search of water. As we crossed, we added our own tracks to the crisscrossed trails and raised small clouds of dust. The air was heavy with humidity, pressing down on us, sticking our clothes to our skin as we entered a leafless forest. The red-chested cuckoo and yellow-billed kite had returned from their migrations, but still the rain held off.

Some hours later, as we walked through dry tufts of red grass, the first crack of thunder echoed across the valleys, and at last the rain started to fall. Within minutes, the heavy stillness of the day was broken as life erupted around us; insects emerged from dry cracks and crevices, the coucals and other birds broke into song.

As we approached a thicket of umthombothi trees, I heard an unusual squealing and hissing on the ground. I thought at first that it was some kind of cricket or cicada, and scanned the area. To my amazement, I spotted two spitting cobras copulating – it was they who were making the sound. They were entwined and facing each other, but had paused, disturbed by our footsteps. They both reared up, facing us, their hoods spread and flattened, revealing the striking bands of gold and dark brown on their chests. I quietened my trailists and gestured to them to move slowly back, lest the cobras spit their deadly venom – which can travel three metres and cause blindness. We watched from a safe distance, delighted when the creatures continued with the act of copulation. After several minutes their movements slackened, and they began to move towards a termite mound and the thicket of umthombothi trees beyond. We watched these potent, graceful creatures winding sinuously together through the wet grass, the glossy golden-brown of their shifting scales gleaming in the rain. They left us with euphoria, our bodies coursing with energy from having witnessed something so spectacular. I smiled at my group, and said, 'Welcome to the wilderness.'

This had happened just three hours into our trail.

Lightning flickered through the dark clouds and thunder rumbled, and the air was electric and alive, as we entered a dense sycamore fig forest. We declared our camp on the soft sand of the iMfolozi River. The Momfu cliff smiled down at us with its face of gold sandstone; on top of the cliff, a troop of baboons chattered and seemed to cackle with amusement at the sight of us. Soon our fire blazed, releasing the sweet scent of umthombothi smoke over the camp as we watched a herd of elephants coming to the dry riverbed to dig in search of water – for themselves and others.

In the morning I reflected on this with the group, talking about how simple acts of sharing resources could reinvigorate our souls and remind us of the essence of ubuntu, which has become so lost in modern life. The elephants have much to teach us – especially our most powerful leaders, who seem to know nothing about sharing. All leaders could learn so much from the elephants, which were using their power not to crush the weak but to create life to share with other creatures.

I spoke of the human hunger for connection with the raw current of life that runs through nature, through the thunderstorms, the copulating snakes, the generous elephants. How this dynamic power reverberates through our bodies and souls and sinks into our blood, reviving the bonds that we have severed by treating nature as a resource to be exploited for profit, instead of a life force that sustains us in every second of our lives.

One of the trailists, Geseko von Lüpke, recorded me as I spoke. I did not know that this simple speech of mine, one of hundreds that I had given to trailists over the years, would mark a turning point in my life. Geseko would send this recording to a radio station in Germany, and tap into a wealth of opportunity that would open the door for me to leave Ezemvelo. Just as the elephants dug holes for the other creatures to drink from, this German was digging a hole for me in the riverbed of my life, which was dry at that time.

After we had seen the snakes, my backup guide, Bheki Thethwayo, had told me that seeing snakes mating portends a new life. He said, 'Sicelo, if you were not married, I would say that it is

Learning about ubuntu from the elephants

now guaranteed that you will get a wife. Are you perhaps looking for a second wife?'

I laughed, saying I was certainly not in the market for a second wife.

'Well,' he said, 'there is definitely a new dawn coming for you.'

I felt a quickening of my heart, for I was yearning for this new dawn. As it turned out, he was right.

My heart knew that I needed to leave Ezemvelo. But my head was cautious, and my feet were slow. There was so much at stake.

Dudu and I had finally got married, and we were expecting our second child. I had many responsibilities, and I knew many people who'd made a decision to leave their jobs without thinking through the consequences, and who were now battling. I had come to Ezemvelo as a volunteer, struggling every day, going hungry for the sake of my passion for the work. Now at last I had the security of a monthly paycheque, medical aid, pension. But my wife had a permanent post as a teacher, and felt that her income would carry us through the dry seasons when my work was slow.

The Menzies project had given me a solid vision of the kind of work I wanted to do, combining wilderness tours with community upliftment. I was inspired by the possibilities of doing this if I ran my own company. That goal pushed me and still pushes me today.

As I was thinking over my decision, my friend Ian Read invited me to do backup for a trail in the Drakensberg. While on this trail, I had a powerful dream.

I dreamt I was in a car accident, with two cars colliding at an intersection. I dreamt I lost my life in the accident and saw my body covered in blood. I took the form of a jackal, and moved away from the accident. Then I came across a collection of beautiful mud houses. Some were half-built, or falling apart, with people trying to fix them. I heard a voice say clearly, 'You still have not finished your mission. You have to go back and finish your mission.'

When I woke, I discussed the dream with Ian. He said that seeing my dead body was telling me that I needed to end part of my life in order to fulfil my mission. He asked what my vision was for my life, and I told him that it was to work with the wilderness and the community, and to help my community and other people. He said, 'Those houses are the lives of the people that are falling apart and your vision is to do that work.'

As I was driving back home, I received an email from Geseko von Lüpke in Germany, saying that they had broadcast my recording on the radio and people had loved it. And that he was organising for me to go to Germany to run seminars with an organisation called Wildniswissen.

I discussed the dream and Geseko's email with my wife. I told her that I was worried that if I left I might not be able to fulfil all my responsibilities to my family. She said, 'But you are not happy where you are, so it is better if you resign, and go and promote yourself in Germany.'

I was so grateful to her for her courage and support, for urging me to make a choice that was right for my life's purpose even if it might make our situation more precarious.

All of these things came together to give me the courage I needed. I drafted a resignation letter. Then I sat on the bed in the house where I was living in the iMfolozi, with the resignation letter in my hand.

Outside my window, the wind was howling mournfully through the branches of an old umthombothi tree, shaking it so that shredded pieces of bark fell to the ground. This tree had been so much part of my life ... for how many hours had I sat, leaning against its trunk after a difficult day at work, allowing my anxiety and stress to melt out of my skin into the skin of the tree? A female nyala came tripping through the thicket, the cream stripes on her golden coat rippling in the dappled light. She walked slowly towards the window, and gazed at me for a minute with her deep, black, liquid eyes. She was hardly a metre away, with only a thin layer of glass between us. It seemed as if she'd come to say goodbye.

Overwhelmed with emotion, I stumbled out to the back veranda to gaze out at my favourite view. Vultures were circling overhead – they'd come to feed on the carcass of an old male giraffe that had recently died. He too had been a frequent visitor to my home in the iMfolozi, but now he was gone.

It was time, I thought. Time to shed the comforts of working in Ezemvelo, as the old umthombothi tree was shedding its bark. Time to accept that this part of my life was gone, as the old giraffe had gone. Time to do what the nyala had come to do. To say goodbye.

I knew that the road ahead would not be easy. That it might bring painful experiences, and mountains that seemed too steep to climb. But I knew too that the fire was still burning and wild in my heart, the passion for the wilderness as strong, and getting stronger, and that I was a warrior not defeated but hungry for change.

I went inside and filled in the date on my resignation letter. I walked out to deliver it to the office, closing my front door softly behind me. As I walked past the old umthombothi tree, the wind sang in its branches.

The winds of change were blowing for me again.

Part Six

———•———

GROWING THE UMKHIWANE TREE

2014 to 2021

THE UMKHIWANE OR SYCAMORE FIG TREE, *FICUS SYCOMORUS*, CAN rightly be called the giving tree. It's one of the magnificent species of the African woodland and savanna. It can reach up to thirty-five metres in height, and spreads its arms much wider than its height to allow all to enjoy its shade. The trunk may grow to two metres across. The bark is a pale, greenish grey, lightly furred on young trees. Abundant clusters of small, round fruits, the size of small plums, grow directly from its branches and trunk, and are pale green to pinkish-orange when ripe.

These plentiful fruits feed an array of animals, including birds, bats, elephants, giraffes, kudus, nyalas, bushbuck, impalas, common duikers, bushpigs, warthogs, baboons, monkeys and bush babies. In former days, Zulu communities would sun-dry the fruits during times of drought, before grinding them into a powder that could be mixed with water to make milk or eaten as a porridge. The sap or bark extracts can be used to treat chest complaints, glandular problems, pharyngitis and diarrhoea. Bark powder may be sprinkled onto burns, fruit infusions are used to treat TB, and a tea from the leaves can help dysentery. The wood has been used as a base block to create fire using rubbing sticks, and to make drums; rope can be made from its bark.

Even when their lives are over, these trees continue to give – their rotting and hollow trunks provide homes and nutrients for many creatures for years after the tree has died. And I have never forgotten that it was an umkhiwane tree that saved my life when I was seven, for it was the stump of this tree that my father beat with a stick to frighten away the lions that were bearing down on us.

It had long been my dream to form my own company, and I had known for many years what I would call it. So when I left Ezemvelo in February 2014, I formed the company uMkhiwane Ecotours, later to be changed to uMkhiwane Sacred Pathways. Its vision was to be as the umkhiwane tree, to grow strong, wide and tall, and to give, and to give, and to give – to my family, to the community and to all who walked with me in the wilderness. A flourishing tree that would inspire people to heal the wounds we humans have inflicted on each other and on the natural world.

I did my first trail as uMkhiwane Sacred Pathways in February 2014, carrying my own grief for my father, who had passed away earlier that month. I got the call to say that my dad was ill when I was near Sodwana Bay with Dudu, finalising the use of a piece of land to build a new home for my wife and children. I rushed to my homestead in Hlabisa. I could see straight away that he was in a bad way. He was diabetic, and gangrene was setting in on his leg. We had not realised this as there was no external wound, and he often had pain in that leg from an injury he'd acquired as an Umkhonto we Sizwe soldier.

I said we should take him to hospital. He didn't want to go, but I explained that we needed help, so at last he agreed. When we arrived at the hospital, the medical staff tried to save him. He was surrounded by doctors and nurses, but I soon realised it was too late. All they could do was try to ease the pain that he was suffering; all we could do was give him the medicine of our love and our time. I hugged my mother, and tried to be strong for my family.

I stood with tears pouring down my face, watching my mother wiping the sweat from my father's brow. My brother, Siyabonga, looked shattered. I could see he was fighting tears. Outside, the last sunlight was disappearing behind the white walls of Hlabisa hospital, as the light was dying in my father.

The ward went quiet for a while, the only sound my mother's soft praying for her husband and the ticking of a dusty clock on the wall. A spiderweb trailed from the clock, as if time was standing still, the hands pointing to four o'clock. But the clock was still ticking, measuring my father's last seconds. I wondered how many tragedies this clock had witnessed.

My dad was looking hard at me and Siyabonga. I could see the realisation in his eyes that his life was going. He called me and said, 'Sicelo, can you please hold the left hand,' and he called Siyabonga, and said, 'Can you please hold the right hand.' As we stood holding his hands, he looked at me, he looked at Siya, and he closed his eyes. I closed my eyes too, and tried to meditate on the colour green. My father loved green ... I thought of the green of new leaves on the bushwillows, green apples, the soft green of summer grass ... I wanted to send him the colour to follow into the unknown world.

I heard the nurse say, 'He's gone,' and opened my eyes. My mother was sobbing, overcome, on the floor, and Siyabonga fell down in a faint. My father looked as if he was sleeping, but his heart was no longer beating.

The nurses could see that I was a little stronger than the other members of the family, so they asked me to attend to the documentation and asked me questions about his personal details, about how long he'd been sick.

A Sunday afternoon in a public hospital is a traumatic setting for the death of your loved one, for the staff are overworked and the corridors crowded with people in need. The staff looked so exhausted that I offered to help them lift the body of my dad as he was moved to a gurney and his body wrapped in white plastic. I helped them push the gurney down the long passages to the hospital mortuary, tears flowing down my cheeks. The nurses walked behind, chatting softly, sometimes giggling – attending to death was part of their job, and for them it was just another day's work.

As we reached the mortuary, a nurse rushed ahead to open the doors. The creaking, clanking sound filled me with dread. We lifted his body and carefully placed it on a shelf. I gently touched his head for the last time and said, 'Dad, I will be back soon to take you home.' I knew that our powerful friend, the buffalo thorn, would help us retrieve his spirit.

I walked outside slowly and looked up at the evening sky, my mind a jumble of memories. I could not believe that this man

My father (centre) with my sister, Makhosi, and Dudu (top left and right), my son Hawu, and my mother (right)

would no longer be in my life. But at least I'd had the chance to say goodbye, to hug and kiss him for the last time, and his suffering was at an end.

A few months after this, on 22 May, my third child and first daughter was born. We called her Ntando, which means 'gift of God', for she truly felt like a gift from the universe. I was still grieving for my father when she came into the world, but as I held her I thought about the healing power of new life that is our only consolation for death.

———————•———————

My first uMkhiwane Sacred Pathways trail was marked not only by my own grief, but also by the trailists' deep historical wounds. On it were two Jewish couples whose great-grandparents had died in the gas chambers and whose other family members had survived the concentration camps, and two German couples whose great-grandparents had been in Hitler's army and had participated in the persecution of the Jews in the 1940s. While the families of the Jewish couples had endured persecution in Germany, on fleeing to South Africa they had been welcomed as white people and benefitted from the racially dictated privileges that were bestowed on them by the apartheid government. So it was that the group carried in their DNA, as we all do, a tangled web of oppression and dispossession, the legacy of forebears who had received and/or inflicted pain.

My backup guide and I also carried layers of persecution and dispossession. And the knowledge that, beyond this paradise, where the animals roamed freely for the delight of the wealthy, lay the hard-pressed pastures of our villages. These communities had been failed, not only by the apartheid government but also by the democratic government that had followed, by corruption and greed. And they had been failed by the park.

Because too little has been done to ensure that local communities benefit from the wilderness in their midst, some residents have embraced the coal mines that are encroaching on the park's borders. These mines have brought jobs. But they have also brought polluted streams and rainwater blackened by coal, making rain tanks useless; explosions and sirens at all hours of the day and night; homesteads cracked by the blasting; community strife and anti-mining activists gunned down or intimidated; grazing lands bulldozed; and livelihoods destroyed, in exchange for employment that will die when the mine is exhausted in ten or fifteen years.

Coal mines are not the only, and certainly not the best, way to uplift the community, and they pose a grave threat to the wilderness in the iMfolozi. While Ezemvelo has invested in community partnerships, I believe that much more can be done. With creative investment from the government and the private sector, it could be possible for the whole area to become a biosphere reserve, with communities benefitting from projects to promote organic and sustainable farming and food gardens, to conserve water resources; to provide pumps and boreholes to save women the drudgery of collecting water; solar water heaters, lights and cookers to stop the burning of coal and cutting down of local trees; composting toilets; and home stays and village tours where crafts could be sold to the tourists. The community has much to offer visitors – cultural tours, beautifully crafted artefacts such as bead and basket work, reed mats, wooden carvings, fresh produce and local dishes, and an authentic, indigenous, rural life experience. With such investment, the community would be motivated to protect the park and its wildlife.

It was my dream to use my company to realise some small part of this vision. I'd met many groups from overseas, such as the people from Menzies Aviation, who wanted to do more than sip gin and tonics in a luxury lodge. This was the first branch of my vision for uMkhiwane Sacred Pathways: to combine the rich rewards of a wilderness experience with the privilege of bringing resources to those in need.

The second branch of my uMkhiwane tree was to open the fences of exclusion, and enable local community members to experience the healing power of the wilderness. My longing to fill this need gave birth to a non-profit organisation under the wing of uMkhiwane Sacred Pathways: uBizolwemvelo (the Call of the Wild), dedicated to finding resources for local communities to experience spiritual journeys in the wilderness.

And the third branch – the trunk, perhaps – of my tree was to heal humans with nature, and to heal nature by making people aware of how much we need wilderness, flowing rivers, clean skies, deep forests and wide savanna, and wild animals roaming free.

On that first trail, we paused at a waterhole to watch the parade of animals as they came down to drink. The dust hung in the still air, stirred up by the hooves of zebras, wildebeest and impalas.

They walked side by side, united in their common thirst, content to share this life-giving resource. Together, they created a river of black, brown, white and tawny grey, each made richer by the contrast with the others.

Later that day, we sat under one of these magnificent umkhiwane trees and reflected on the experience of watching the multitude of animals cooperating and intermingling peacefully as they came down to drink, on how seeing this had opened the doors to the possibility of other ways of being in the world. It was a warm, dry, winter's day, and everyone had finished their water. We had only one full bottle, which I kept for emergencies, and we agreed to share it. I watched the bottle as it was passed from hand to hand, listening to the glug-glug of water as each drank in turn. We had come together, each carrying different burdens of our sad, uncanny history; but, like the animals, we were united in our thirst and in this simple act of sharing water. The shared water bottle invoked a vision of a kinder world, and became an oasis for our bodies, our hearts and our souls.

I lay back against my backpack, lifted my eyes to the dense green canopy of the tree against the limitless blue of the sky, and gave thanks for this moment in this place. My uMkhiwane Sacred Pathways journey had begun, and I needed now to trust the pathway I had chosen – and follow wherever it led.

———

When my father held my hand as he lay dying, it had felt like his blessing. But despite this, it was tough to get my company going.

I had no shortage of people who wanted to do trails with me, but I was struggling to get permission from management to lead trails in the park. I was given no reasons for this, but my requests were often ignored. Or I would be given permission for a trail, but told that I was not allowed to lead it, even though many freelance guides were allowed to lead trails. I could still lead trails under the Wilderness Leadership School, but their fees pushed up the cost for participants and limited my earnings to a basic subsistence.

This was a stumbling block I'd not anticipated. Luckily, I had purchased a minibus and managed to build up a business transporting tourists from the airport to the park, but transporting tourists was not my vision. I spoke to many on these trips, and they became interested in doing trails with me. I did some in the Drakensberg, but I was having to turn clients away who wanted to do trails with me in the iMfolozi. And I was not making the gains I wanted for my community.

At that time, partly thanks to assistance from my old friend and helper Richard Knight, I was taken into a Department of Tourism incubator project to promote small black-owned tour guide businesses. This offered me various platforms to help market my tours. With their support, I went on a sponsored tour of Swaziland, Mozambique and the Seychelles to experience the lodges and get an idea of 'what tourists wanted'. It was interesting going to these places, some of which were very beautiful, but I quickly learnt that this was not the sort of tourism I was interested in promoting. It seemed to be all about gala dinners, nightclubs, cocktails at the bar, being taken from one luxury setting to another. The only real contact I had with the places was when I managed to escape and chat to locals about their lives – that was how I learnt something of the devastating civil war that still haunts Mozambique.

I was grateful for the opportunity, but came back even more resolved to develop something different with uMkhiwane Sacred Pathways. I'd met so many tourists in my work who wanted to go on true journeys of the soul. I knew that there was no shortage of clients wanting meaningful interactions with the nature and people in our country. I just needed to keep finding ways to grow my tree, and to trust that the growing network of people who shared my vision would help me.

In 2015, I was given an opportunity to grow one of the uMkhiwane tree branches that was close to my heart: to take my local community members into the iMfolozi wilderness. I set off over the hills of Impila Encane with eight young adults and one elder from my community, and Richard Mchunu, my backup guide. The hills are named after the impila shrub (ox-eye daisy) that flowers on the ridges in early summer. A spreading umbrella thorn dominated the horizon with its distinctive outline, but the grassy slopes around us seemed, for the moment, empty of animals. We could hear the rumbling of an impala ram seeking a mate, but we could not see it. We followed an animal path down into a forest, beneath the Momfu cliff with its yellow sedimentary rocks. It was autumn, and the leaves of the bushwillows and camphor trees were turning yellow and beginning to fall. The path was littered with dung and fresh spoor, but no animals could be seen.

There was another invisible presence on the trail. Like the animals, she could not be seen, but her spirit was strongly with us.

Fiona was the wife of a good friend, Trevor. It was only a few weeks before that I had spoken to him on the phone, his voice shaking like the reeds in flooded rivers. My wife has died, he said. Fiona has died. I tried to console him; I asked to hold a meditation for her soul. But he had lost his life partner to an illness following cancer, leaving their three young children motherless. What comfort could I offer? Hearing the depth of my friend's sadness paralysed me, choked my words of condolence.

Trevor was a general practitioner based in the small Zululand town of Nquthu, whom I'd met on a trail ten years previously. He was a tall Irishman, full of energy, with the swashbuckling good looks of a 'Marlborough man' – piercing eyes the blue of Mediterranean waters, flowing hair tied back in a ponytail that swung from side to side as he walked. His great-grandparents had emigrated to America to escape the potato famine, but his parents had returned to Dublin, where he was born. Like his great-grandparents, he was hungry for adventure – he'd come to South Africa as a medical intern, and never left. When I met him he had spent several years working in the heart of Zululand, and had a deep knowledge of the customs and the language. After we did a trail together, he invited me to spend a few days at his farm, horse riding and tiger fishing in the Pongola River and Jozini Dam, and we became good friends.

Fiona was a powerful woman, supporting Trevor in his work and working tirelessly for the community herself, initiating projects such as a small children's home. Trevor and Fiona came on trails almost every year, and it was Fiona who had suggested that this year they'd bring four from their family and subsidise four trailists from local communities to go along with them.

As I lamented with my friend, Trevor told me that he would like us to conduct the trail in honour of Fiona, instead of cancelling it, and give four additional members of my community the opportunity to go in their place. I was so moved by this gesture. The world truly has good people in it.

The zigzag path up the Momfu cliff was steep under the late summer sun. But the reward was worth it, as we came over the rise to the full beauty of the iMfolozi River laid out before us. The silver surface was broken by a lone waterbuck, its curved horns silhouetted against the light. A brown scrub robin called from the bushes, its high, sweet melody seeming to celebrate the graceful lines of the antelope below. I noticed that the footsteps behind me had ceased, and turned to see my companions standing in silence, awed by the beauty of the scene before them. When they saw me waiting, they moved to catch up.

'Did you see something?' I asked.

'Nothing and everything,' a woman answered softly.

She did not need to say more. I could see in their faces that they had been touched by the splendour of nature, that unique moment when you are oblivious to everything but the inexpressible beauty before you.

We continued up the path, over rocks exposing red iron through the withered grass. I explained that this rock could produce iron if smelted by skilled blacksmiths – this was how our ancestors had created spears for the soldiers, spears that were used in battles on all these hills around us. I pointed out the iNqabaneni hill nearby, and narrated the story of a fight between Shaka and Zwide. During the bloody battle, Zwide refused to come down from the hill to engage with Shaka and his waiting warriors – this was how it came to be named iNqabaneni, the hill of refusal.

The sun was warm, and we were tired after our climb. We dropped our packs, took off our boots and lay back, lifting our faces to the sky. I could hear the crackling of vegetation and a rumbling of elephants in the distance, and knew they were approaching the river for their afternoon drink, but they were still some way off. As my mind drifted into waking dreams, Richard Mchunu, who had been keeping watch, called, 'Vukani! Wake up! Elephants are drinking; you came here also to see these flowers.'

One of Richard's favourite sayings was 'izilwane zasendle ziyi-zimbali ezihlobisa izinhliziyo' – wild animals are like flowers that decorate our hearts. Hence his reference to the elephants as flowers.

Richard was an old guide who'd been with Ezemvelo for thirty years – he used to work with my dad at kwaMakhamisa section, so I'd known him as a child. On this trail I was his leader, but his knowledge of the bush and the reserve was prodigious. Although he was the backup, he was often leading from behind, and I always appreciated his insights and information. He cared deeply for the dangerous creatures, and all other creatures, in the wilderness, and always insisted that they be treated with caution and respect.

As if the elephants had brought the wilderness to life, the

landscape was suddenly teeming with animals. A bachelor group of buffaloes emerged from the bush, and came sparring and jostling into the river; a family of giraffes glided by in slow motion, their heads nodding above the trees as they walked. Blacksmith lapwings swooped above the river, their calls like the chink of metal against metal, their black-and-white plumage reflected in the water.

'More flowers, uMkhulu!' a young man declared to Richard, pointing at the buffaloes. Richard laughed, and nodded his head. His weathered face was alive with delight. I knew that he was resonating with my own passion for sharing the wonder of these 'flowers' with the young adults from our community.

The sun was beginning to sink behind the Lubisane hills – it was time to head to our campsite. We made our way down the slope, and along the river towards the rocky shelf where we would spend the night. As we approached the river the elephants caught our scent, and stopped playing and drinking. The matriarch raised her great trunk in the air to check our scent, then calmly dropped it to fill with water, before lifting it to her mouth to drink. It seemed to be a message from her that we could pass by without disturbing the herd. We walked past slowly and cautiously, our own footprints mingling with the elephant's huge round prints in the golden river sand.

Soon we had a warm fire crackling, and a kettle bubbling. The reeds danced with the evening wind and the ripples on the iM-folozi glittered in the evening light. Nearby, a family of three elephants – a mother and two young bulls – and a subadult female were browsing peacefully. They were so close we could see their thick eyelashes and smell the grassy sweetness of their breath.

'Indlovu iyabhodla,' Richard laughed softly – the elephant is belching.

I felt a strong sense of being welcomed by these creatures as we shared this simple intimacy by the river, each enjoying the companionship of our own species and others. I thought of Fiona, of how

Richard Mchunu

she'd had the same generosity of the stately elephant matriarch at the river. Fiona brought kindness to the world, making life easier for so many. And even after her death her generosity lived on, in this last act of kindness that brought this beautiful experience to eight community members, and to me.

The lives of people whom Fiona had never met would change forever because of this. I thought of what Trevor had said, that he wanted this trail to go ahead in her honour, and it felt indeed as if her spirit had been with us at every step – in the call of the scrub robin and the lapwings, in the graceful curve of the waterbuck's horns, in the sweet grassy breath of the elephants.

Inspired by this trip, I strengthened my efforts to raise funds for uBizolwemvelo. By 2016, with the help of my friend Bridget Pitt, I'd raised enough money through crowdfunding for a pilot trail.

At last I had the funds, but again I ran into problems when trying to get permission to do this trail in the iMfolozi. I decided to use this obstacle as an opportunity and take the group south for a trail in the Drakensberg mountains.

There were seven teenagers, both boys and girls, and an elder,

Baba Hlope. Many had never travelled more than a few kilometres from their village. Now, they were travelling six hours away into the mountains. As we set out, I remembered my first trip away from my home to Durban – how different our destination now. Instead of towering buildings, we would see towering cliffs; instead of rubbish flying around, we would see the Verreaux's eagles and lammergeiers (bearded vultures) that ride the thermals above the high slopes.

We drove along the dusty roads near my village until we joined the R618 that winds through the Hluhluwe–iMfolozi Park. This corridor separates the iMfolozi and Hluhluwe reserves, but animals can move freely across it. The hills of kwaHlaza were blanketed in trailing mist; the heads of a family of giraffes protruded above the fog, watching us curiously. A soft summer breeze swirled through the open windows of my car.

We continued through the rural villages and past the sprawling Somkhele opencast coal mine, a gigantic crater gouged from the green hills that cast a menacing shadow over the nearby home-steads. The bushes beside the road were choked with black coal dust. Soon after we joined the national road (the N2), we began to see the military rows of the eucalyptus plantations. I explained to the youngsters that these trees were not indigenous, and caused much damage to indigenous plants as they are greedy for water. In between the eucalyptus were sugar cane plantations, wave after wave of green sugar cane rippling with the soft sea breezes. The youngsters were amazed when I called this lush vegetation a desert, but I explained how a monoculture crop like this kills all other life. The soil is depleted, forcing the farmers to use more and more fertiliser to force more crops out of deadened soil; the rivers are poisoned by this fertiliser, leading to algae growth that suffocates aquatic life; the insects are killed by pesticide; no birds flit between the waving fronds.

I was glad for the opportunity to speak of these things. These youngsters only ever heard a different side of the story, that sugar cane and eucalyptus brought wealth. But this wealth is like cheese in a mousetrap, for it comes at a heavy price. I felt such gratitude for this opportunity to crack through the shell of ignorance that confined their horizons. The dream of uBizolwemvelo had been in my heart for such a long time, and I felt gratitude for all the help I had received from those who'd done trails with me, or supported our crowdfunding campaign, to make this possible.

The landscape shifted as we turned inland to the Natal Midlands, the domain of huge dairy and beef farms. We drove through emerald-green pastures dotted by patchwork black-and-white cows – their heavy udders, dripping with milk, astonished my community members, who were used to our free-range cows at home that are allowed to keep their calves with them. Gradually this, too, changed, as more belts of coastal scarp forest appeared, lining the road with impenetrable vegetation. We crossed the uMkomazi River, gashed with silver rapids, the dark slate and sedimentary rock gleaming like black satin beneath the water. The purple line of the uKhahlamba mountains rose up like a dream on the horizon.

As we came through the town of Bulwer I smelled rain, and lightning began flickering on the still-distant mountain slopes. Soon, a heavy downfall engulfed us, blurring the mountains behind a cascade of raindrops. The dirt road became a slippery river of mud. A black-bellied bustard glided smoothly beside the road, coming to land on the soft clover pastures of a dairy farm. As the rain thinned out, we at last saw the sign – Welcome to Cobham – and, shortly afterwards, pulled into the parking area of our base camp.

I stepped out onto wet grass, cold under my bare feet. A chilly mountain breeze blew through my threadbare T-shirt. The group clambered out after me and began dancing, ululating and singing in the rain. Two girls prayed to uMvelinqangi, the creator of all things for the amaZulu, giving thanks for the safe trip; some of the boys were uttering King Shaka's praise name, Udlungwane lukaNdaba, giving thanks to all the ancestors. I knew that we had arrived in a place that offered us not only mountain heights, but heights of self-reflection and joy.

I pointed out Hodgson's Peak as it emerged above the wreaths of mist across the escarpment. I could smell and feel the youngsters' hunger to savour every moment of this opportunity to experience these mountains, for I had felt that hunger myself when I was their age. We laid out the equipment and I explained how it should be packed as the sun emerged through the clouds, cast a golden light through the yellowwood forests and threw long shadows across the green slopes of the valleys. The light played on the face of the Ndlomvini mountain, catching its golden hues.

We set off the next day on our three-day trail, aiming for the cave where we were to spend the first night. My boots welcomed the mountain path that took us through straggling clumps of umtshitshi (oldwood) bush. In the blue distance, the imposing 'barrier of spears', the uKhahlamba-Drakensberg escarpment,

The rocks are our teachers

brooded over us. A jackal buzzard circled in the sky above us, its hoarse, bleating cry mimicking the call of a black-backed jackal. We walked in single file. No one spoke. There was just the heavy breathing of those unused to the altitude as we climbed the steep path, deep into the mountains, far from civilisation.

The sound of the Pholela River calmed my thoughts, and soon we rounded Boundary Rock, so called because it marks the boundary between the sheltered lowlands and the high wilderness. The huge, grey boulder jutted out from the slope, like the head of a whale emerging from the waves. On our left, Siphongweni Rock loomed high, forbidding and inaccessible, but this was our destination for the night. The late-afternoon sunlight illuminated the valleys on the far side of the river below us, coating the green hills with a golden glow. A lonely giant tree fern caught the sun, each green frond a spectacle of light. We passed a small stream, cold sparkling water running over dolerite rocks.

My legs ached and my shoulders were numb from my heavy pack as we scrambled up the last climb, a rocky, slippery ascent with a drop of six metres. I reached the top, swung my pack off my shoulders, and sat down on it to await the others. As each young-ster made the climb, out of breath and sweating but with a happy face, I welcomed him or her into the wilderness. I wanted them to bring themselves into the present before we ventured into the cave. I invited them to lay down the heaviness and tedium of their daily lives on these generous slopes of ezintabeni ezimnyama (the black mountains). The wind blew with a ghostly whistle through the stiff-limbed sugarbush trees.

We climbed the last few metres to the cave, which is formed by a hollow under a massive boulder. The main entrance was hidden by shrubs; a small shallow stream flowed through the middle. The smell of fresh izimpofu (eland) dung and porcupine droppings reminded us that this cave was also used by other creatures. By the time we had boiled water on the Primus stove, the sun had set.

Young hearts opening to the power of the uKhahlamba-Drakensberg

Warming our hands on cups of tea, we clambered out and sat on the rocks, watching the darkness creeping over the escarpment, enshrouding the lakes and valleys.

We spent the next day wandering over the high slopes. My young charges were fascinated by the sugarbushes that dotted the mountainside, especially when I explained that these plants rely on fire to regenerate and germinate their seeds. We spoke about how beauty can come from hardship, how flowers may be born of fire. It is like us, one boy said. Our Zulu people suffered the fires of the white settlers, and the white governments, but we endure and create beauty from that hardship.

We returned in the afternoon to explore the Siphongweni Khoe and San rock art site, further up the ridge from the cave where we were sleeping. Here, a broad shallow cave has been formed by a long overhanging rock ridge. Every possible surface is alive with ancient paintings, some dating back three thousand years. We stood in the warm afternoon, examining the delicate lines of the eland, of the hunters with spears, of shamanistic figures, half-human, half-animal, or with lines radiating from their heads. Later paintings depicted men on horses with guns, a painful record of how the lives of these people were shattered by the arrival of white settlers.

The youngsters were amazed to hear that there had been people in these parts long before our own Nguni ancestors arrived some six hundred years ago. Like most Zulu children, they had been raised to believe that the amaZulu were the first people to walk these lands. They were even more amazed to discover that many of our rituals and medicines had been learnt from the San people. Their idea of these people was very limited: they saw them as some kind of odd wizard, or not quite human. They certainly had no idea that there were still San people living in parts of South Africa and Botswana. I could see, as we talked, that their whole concept of themselves and their history was being challenged.

We spoke about how the hunter-gatherer San did not understand the concept of owning cattle, but lived harmoniously with the first Nguni pastoralists. But when the British colonists arrived, they hunted out many of the wild animals that roamed the plains. The San turned to hunting cattle, because wild game was scarce, leading to conflicts with the Nguni people as well as the white settlers. They fled into these mountains, and documented their trauma on the rock walls of this and many other caves. The youngsters were shocked to hear that the white settlers had hunted the San like animals and, up until 1906, were paid a reward by the government for every San person killed.

We spent the afternoon sitting quietly in that place, listening to the murmur of the spirits of those who had gone before us. We spoke about the paintings, how the San people made paints from ochre, ash, charcoal and blood; about the messages behind the marks on the rocks; about the artefacts and tools they used; about the food they ate; about their wisdom and knowledge of medicinal plants, how this knowledge had been passed on to the amaZizi and later to the founders of the Zulu nation. Above us huge rocks towered, sentinels guarding this refuge of the ancient ones, the first peoples to walk this land.

As the sun sank behind the mountains, we made our way down the slope to our cave for the night. We ate our supper, listening to the plaintive calls of the jackals, the sorrowful calls of the dassies as they scurried to conceal themselves in the cracks and crevices of the rocks. I hoped that this cave could offer these young people a refuge, as the rock crevices offered the dassies a refuge, as these caves had offered the San people a refuge. That this refuge could be carried home in their hearts, giving them a place to hide and to regenerate their souls; giving them the strength they needed to endure the fire, as the sugarbush did, and to create something beautiful in their lives.

Some weeks after this trip, these young people came to me and told me that they had been exploring the hills near my village. They said they found about eleven caves, some with San paintings. I told them not to tell anyone, because people might vandalise them. But it pleased me that they had found refuges near their homes,

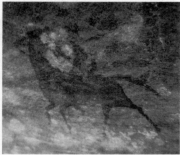

The spirits of those who came before us

The earth is our refuge

refuges where the spirit of the San people still endures despite centuries of dispossession, enslavement and genocide. Refuges, in the mountains and in their hearts, where their own spirits could endure, and bring them comfort.

———•———

Two years after my father died, my family performed the ihlambo, a ceremony to cleanse the family of mourning and grief, and welcome my father into the realm of the ancestors. This ceremony would also mark the moment when my older brother became the official head of the household.

An important part of the ceremony was the ritual hunt. My brother carried our ancestral spear, which had belonged to my father, and was broken when he died, according to custom. Now, the head of the spear was bound onto a new handle, and the spear was taken up by my brother as the eldest son.

The hunt is a symbolic celebration of the oneness and unity of all things, through the act of taking a creature of the bush and uniting it with the homestead by bringing it home as food. At midday, the hunting party washed themselves in the river, and returned to the homestead, with two common duiker and a rabbit. I greeted them with other family members, singing our family's ihubo, a sacred song that speaks of the crack of dawn and the new light coming, of the protection of ancestral spirits. We took the meat from the hunt into our ceremonial rondavel to be presented to the ancestors.

In the evening my mother was taken down to the river by female family members to be washed and dressed in new clothes. Her black mourning clothes were burnt.

My father was now finally at rest, and could take his place with the other ancestral spirits. I knew that my relationship with him held both light and darkness, love and betrayal. But I was grateful for this ceremony which helped me to process my feelings. I hoped that with the blessing of his ancestral presence I would be able to resolve some of the difficulties facing me, and achieve my life's vision of growing the giving umkhiwane tree.

———

One night in 2017, I woke to the soft patter of rain on my sleeping bag. Half-asleep, I crawled out and began pitching the flysheet. I called to the woman sitting by the fire, 'Make sure it doesn't go out.' She looked surprised but obediently added a couple of logs. There was an eerie 'whoap, whoap' sound echoing through the trees ... what animal could that be? I reached for my rifle, but it wasn't there.

A moment of panic, then I realised.

I wasn't in the iMfolozi. I was in Germany. The 'whoap, whoap' wasn't an animal – it was the whine of the huge wind turbines slowly turning nearby.

For me, waking up with rain in the night, surrounded by white people, was the normal scenario of a trail in the iMfolozi. But as my reality dawned on me, I felt tears trickle down my cheeks, mingling with the rain. I was crying because I was homesick. I was crying because I was in Germany, thousands of kilometres from all that was familiar, but also because … I was in Germany!

My trip had been sponsored by Wildnisschule Wildniswissen and made possible by Geseko von Lüpke, the German who had recorded me speaking to trailists when I was still at Ezemvelo. Geseko is an author, journalist and leading thinker in the fields of ecopsychology and deep ecology. His philosophical belief that the heart of the ecological crisis lies in the separation of humans from their natural environment resonates with my own mission to enable people to be the path of wilderness and not just to walk it.

On the day of my departure I flew from Durban to Munich via Dubai, my first experience of travelling beyond our borders. I'd been invited to do a six-week tour, which would include trails in the forest, walks, storytelling, seminars and an international wilderness guides conference. As the earth fell away beneath the aeroplane, I was terrified of what would be expected of me in the weeks ahead. I'm not a wilderness therapist, I told myself. I'm just an ordinary wilderness guide who leads from the heart. What can I say to these people?

As rural black children, we could never have imagined travelling across the seas to impart wisdom to white people. Yet I had been invited to this country because my insights and philosophy were valued. Breaking these barriers filled me both with pride and with terror.

I emerged at the airport to be greeted by a tall woman with a strong, open face and welcoming smile. 'Are you Sicelo?' she asked.

'I am Barbara Deubzer – I will be looking after you.'

Barbara made every effort to help me feel at home. Although not much of a meat eater, she offered to stop at a shop to buy me meat, as Geseko had told her that Zulus only eat meat – I assured her I was happy to eat whatever she was eating. In the following days, we became good friends.

I had a few days to settle in, before conducting my first seminar with Barbara. This took place over a weekend in a forest outside Schondorf, a small lakeside town a few kilometres from Munich, within sight of the towering Austrian Alps. Barbara and I arrived the day before to set up the big wigwam where we would conduct some of the sessions.

From the moment Barbara discussed the programme for the weekend with me, I realised this would be something new. Every minute of every day, from seven in the morning until nine thirty at night, was crammed with activities. On my wilderness trails, I do everything I can to help people break away from the tyranny of time. I may always follow a pattern, but it is a pattern created by the wilderness, and I follow it as the vultures follow thermals.

But here I was in Germany, preparing to meet my first group, with a schedule in my pocket and apprehension in my heart. The group arrived promptly the next morning at seven thirty, just after Barbara and I had swept out the tent, got a fire going and made tea. I looked at the group of fifteen or sixteen people huddled in their dripping raincoats, warming cold hands on steaming mugs, eagerly awaiting my words. I told myself that whoever had invited me here had done so because I had something to offer, something that they valued, that these people had come here that day because they had been inspired by hearing me on the radio. I had to trust that.

We went through the weekend under a steady drizzle of freezing rain. For every minute of the weekend, I was surrounded by people, right under my nose, awaiting my next input or instruction. I found it exhausting at first, but they were so generous, so appreciative,

Building ubuntu in Germany

and seemed to be genuinely moved by my words. As the hours passed, I relaxed into it, and in the seminars and sessions that followed I learnt to create breathing spaces by inviting people to reflect in solitude, or go on silent walks.

After three weeks in Germany, I met up with Geseko at the International Gathering on Rites of Passage in the Wilderness in Königsdorf, Bavaria, with over two hundred wilderness guides from about thirty nations, as well as other guests. I was alarmed when Geseko asked me to address the conference. I'm used to standing in front of eight people. At some of the seminars I'd conducted there had been twenty. But to talk to a room of five hundred …! As I walked to the stage I could hear my ears ringing and my heart pounding so hard I thought I might fall down. The room was so quiet. I took the mic and said, 'Hello …'

Silence. I could hear a fly buzzing. Someone dropped a piece of paper, and I heard it slide to the floor.

At last I found my voice. I knew from what I'd seen that pure wilderness was hard to find in Germany. But I encouraged people to find a sacred site that was meaningful to them, even if it was just a single tree. If they could walk in the forest, they should find a favourite tree where they could find some quiet and solitude, where they could commune with the tree, or the earth, and share their problems, and find the solace and comfort that nature offers. Afterwards, many people came to tell me that my talk had touched them, and many asked me about organising wilderness trails with me in Africa. A political leader told me that my message of ubuntu was so important for Germans, and thanked me for helping her country.

The day taught me yet again how much people all over the world hunger for connection, with each other and with the earth.

———

The leader thanked the glowing stone before him, and sprinkled it with sage. The leaves curled and crackled, sending out small sparks. A sharp scent filled my nostrils, reminding me of the impepho we burn at home in our own cleansing rituals. The leader poured water on the stones, releasing billowing clouds of steam. The heat was almost unbearable. My face was dripping with sweat. Through the hiss of the water on the hot rock we could hear his soothing voice: 'This stone represents how to be a father.'

It was my first experience of a sweat lodge, which was being offered at the International Gathering in Königsdorf. I'd learnt about sweat lodges from people like Meredith Little and through my readings on Native American philosophy and cultural practices, and was keen to experience one. Geseko was also eager for me to learn more about them, with a view to possibly running sweat lodges in South Africa.

Ukuqguma, the act of steaming to purify your body, is a common Zulu practice. I used to do it often, especially as an adolescent. My grandfather would steam me with lemongrass to control body changes such as pimples on my face. But this sweat lodge was much more intense.

There were a number of lodges with different intentions on offer for men and for women. As I considered which lodge to choose, I realised this was a well-timed opportunity. The trip had been a great success so far, but at home I was suffering because of my difficulties in getting permission to do trails in the iMfolozi, which was seriously hampering the growth of my uMkhiwane tree. I felt trapped and desperate, and my frustration was taking its toll on my family. Rather than channelling the energy of a calm matriarchal elephant, I was like a young bull who had been evicted from the herd, meeting my problems with anger instead of thinking about how to get around them. I was especially concerned about the tension that was growing between me and Dudu. She had been so patient and supportive, but I could see my frustration was beginning to wear her down.

So when I was offered a sweat lodge with an intention of purifying my essence as a father, a husband and a man, I knew it was what I needed.

As I approached the fire, I was astounded by the size of it, by the high bolts of flame leaping into the sky. Lodged in the coals at the base of the fire were a number of small boulders, the size of an adult human head, some already glowing red from the heat. The lodge was a few metres away – a beehive-shaped structure draped in blankets and hides, like a traditional Zulu grass dwelling but smaller.

Tending the fire was a thin, blonde woman brandishing a long-handled pitchfork – these are used to extract the stones and have to be particularly long, as the heat makes it impossible to get close to the flames. The leaping flames, the red glow on her face, the pitchfork, all brought to mind the pictures of hell that had been painted by our preachers, and added to my trepidation. My apprehension earned laughter from my German companions who teased me for being brave enough to face lions and elephants yet being daunted by a bonfire.

I skirted the fire, and followed Geseko and the other men through the low entrance. Some were naked, others wore only underpants. Inside, we crammed together in the small, dark space.

Our leader gave a briefing about the intentions of the sweat lodge, and explained that each stone brought in represented something different, such as inner riches, prosperity, cleansing of the soul. After thanking the stone and burning the sage, he poured water onto it, cloaking us in hot, dense steam.

'Being a good father takes courage. As a father, we provide for and protect our kids; we feed them with food and knowledge, we share love and respect with them.'

The heat penetrated my body, sinking down to my very core. I thought of the hardships that have come to the world through poor fathering. A pack of wild dogs or wolves work together to sustain one another, to care for the sick ones, the weak and the old ones, to care for the cubs. But in our human world, that care is melting away.

I thought of the young people in the village whose fathers had left, leaving their mothers to struggle; the schoolgirls who'd fallen pregnant, repeating the cycle; the fatherless boys who'd turned to drugs or alcohol.

I thought of my own father slinking into town with his pension to sell off my future for women and drink.

I thought of my father beating the umkhiwane stump and facing down the lions to save my life.

I thought of the light in his eye as he held my hand for the last time.

The darkness in the hut was thick, elemental. I could not see my hand when I held it to my face. It lifted briefly when the blanketed doorway was opened to take in another glowing stone. The heat, already intense, grew worse. The stone cast a red glow on the faces around me. The leader told us that this stone represented fertility; that the heights of life could be not achieved by materialism, but should be based on happiness. That happiness is the foundation of everything, that happiness attracts success, health and prosperity. Happiness is the mother of strength and courage and can lead you to wisdom. His voice rang out through the hot darkness, pure and clear, so that I stopped feeling the heat and felt only the wisdom of his words running through me like a river.

As he poured water on the stone, the elements of fire and water coursed through us, purifying our souls. We had been an hour in the lodge, and all were soaking with sweat. The man next to me held my hand and shouted German words that I could not understand. I heard my own voice bellowing as the heat hit my face. Another man behind me howled, his hoarse tones bringing to my mind the haunting calls of the black-backed jackals in the Drakensberg. Some were crying, singing, shouting, the darkness a cacophony of sounds as we became one animal, transformed by the heat into one spirit, one body.

I felt myself enter a trance, with images of fire and water running through me: the imphepho burning in my homestead as we summoned the ancestors, the night fires on my trails, the white-hot lightning crashing about us as we crouched in the cave with baboons, the deluge of rain, the running waters of the iMfolozi, waterfalls cascading down mountain rocks.

I thought of the sugarbush, the flowers born of fire.

Fire is life, water is life.

Fire is death, water is death.

I thought of the fathers of my family: my own, the grandfathers I had not known, the fathers before them, going back and back in time. I thought of myself, the tender fragility of my children when I first held them, the trusting grasp of their fingers when they slipped their hands in mine. I thought of all those who had been fathers to me.

My family: Ntando, Dudu and Hawu

I saw my grandfather holding the glowing spear in the darkness of the hut where I had undergone my transformation from childhood to manhood, and felt some new transformation. Something that could not be described burnt to death in me. I watched it glow, then crumble into white ash.

The leader continued with his teachings, and the sounds quietened to soft grunts, breathing and weeping. As he conducted the closing ritual with the last stone, I came to realise that we all needed to give ourselves this opportunity not only to melt down our sorrows, but also to rebirth ourselves. The lodge had been as a womb, a womb in which my soul had danced with the flames and been washed in steam. I crawled out, blinking in the light, after three hours of heat and darkness. I looked up at the poplar trees, their leaves shimmering in the wind, at the shafts of sunlight breaking through the clouds, dancing through the branches. A soft breeze cooled my smarting skin.

I had left something behind, some dark fury that had not served me well. I gave thanks for the sweat lodge, and walked uncertainly towards the light.

In 2018, I returned to Germany for a three-month trip, made possible this time by Markus Hirschmann of an organisation called Institut für angewandte Kulturforschung e.V. (ifak), which was devoted to promoting cultural understanding across the world. My trip was part of a project called Bildung trifft Entwicklung (BtE) (education meets development), which works with the Wildnisschule Wildniswissen in using innovative educational methods with school learners to promote global learning, development education and environmental education. The trip took me to Austria, Bavaria, Nuremberg, Munich and here, to the Wildeshausen, a forested estate near a small town close to Bremen in the north of Germany, where the Wildnisschule runs trails and nature immersion experiences.

With Markus Hirschmann

I came to Wildeshausen a day early, to set up camp and give myself some much-needed solitude to prepare for the programme ahead. The forest was dense, carpeted in wildflowers and moss, but the rumbling sound of traffic nearby reminded me that this was Europe, where signs of human 'civilisation' are never far away. I searched for a flat spot to pitch my one-man tent beneath the poplar trees. The tall oaks were beginning to drop their leaves, and a light drizzle was falling. I erected the tent, and unrolled my old sleeping bag on the groundsheet. As it opened, a strong smell of umthombothi smoke invoked a wave of homesickness. In the seams of the bag were a few grains of sand from the banks of the iMfolozi River, as if a part of the river had chosen to travel with me to remind me of home. I felt such longing for my beloved iMfolozi and the creatures who live there, intermingled with gratitude for this opportunity once more to share what I have learnt in the wilderness.

After pitching the tent I took a walk under the canopy of trees, my bare feet on the damp, grey soil. It was warm, despite the inter-mittent drizzle. The forest floor was carpeted in sweet-smelling flowers that I couldn't identify – red, yellow and white, their petals opening to reveal depths of colour inside. It was so strange to be among unknown plants. A movement caught my eye – a badger, similar to our honey badgers, emerged from the undergrowth, scratching at the soil and nibbling something on the ground. I crouched down on the damp ground beneath a sturdy oak tree to watch it, taking in the black-and-white markings on its face, its tongue flickering around the food it found, the flies bothering its eyes, the neat round ears twitching. The badger spotted me and paused, watching me for some time, then continued its search, darting into a burrow then reappearing and pausing to check on me before continuing its foraging.

I felt that leap of connection, as we watched each other – a human from a faraway country and this elegant wild resident on

With the team and schoolchildren at Wildeshausen

home ground. My heart filled with peaceful energy and gratitude, for this was my first encounter with a wild animal in a European forest. The meeting with this gentle, engaging animal somehow leapt the many physical and metaphorical barriers I had crossed to be there. I sat listening to the endless song of the crickets harmonising with the deep stillness of the forest, and thanked the little creature for knitting me back into the heart of nature.

The group arrived the next morning, about thirty sixteen-year-olds from schools around Göttingen. We sat in the big communal tepee and I spoke about ubuntu, how it binds us all to live harmoniously with other humans and all creatures and life forms in the natural world. I knew that it was not my words that spoke loudest, but the voices of the trees, the birds, the streams around us.

The fire was blazing, the dry pine crackling and sending sparks into the air. The blaze threw our long shadows onto the canvas walls so that we seemed to be encircled by a group of silent phantoms. The tepee was quiet, the eyes of the children fixed on mine, their minds seemingly open and soft like fertile soil. Young

We are at home everywhere on earth

minds are receptive, but there is also much that is toxic in the world for them to receive – it was wonderful that this organisation was opening a gateway for them to hear the wisdom of the wilderness.

It was inspiring and rewarding to work with this group, but after a few days I began to crave solitude. I missed home, and walked through the beautiful but foreign trees scanning the undergrowth for something familiar.

One day I set off for a walk on my own, but soon had a following of about ten people. I asked if we could walk in silence. As the path zigzagged along near the river, I was surrounded by foreign plants, every leaf and twig reminding me that I was a stranger. But gradually the trees changed from the tall, upright oaks and poplars to ones more gnarled and twisted, much like those I knew from Africa. The river shifted: once slow and deep, it was now a stream gurgling between rocks. I had a compelling sense that I had been there before – and suddenly realised that the landscape resembled, exactly, the banks of the Intshevu Stream, that beautiful but dangerous stream near Masinda Camp which is often filled with crocodiles at the point where it joins the iMfolozi. It seemed so remarkable that I had found my way, in a sense, to this beautiful but dangerous stream by coming to Germany. Standing there in that moment, I felt a deep sense of belonging, as if that small pocket of forest was a portal connecting my African home to Europe. I had travelled so far, yet here was my home, right before my eyes, in the palm of my hand.

There are many barriers in the world, increasing in number as countries try to keep refugees and immigrants out.

Yet we are all creatures of the earth. Everywhere we set our feet is our home.

⸻

Every grandmother who has grown vegetables will tell you that nothing will grow in depleted soil. Just as we need to feed the soil to grow vegetables, my work over the years had shown how hard it was to grow the souls of our communities without satisfying their need for food, housing and health. The wealthy people I'd met in Germany or on trail were hungry for connection with nature, for a deeper knowing of the world. But people in my community were often overwhelmed by more pressing needs. It is hard to reflect on your connection with the living world when your stomach is empty and your mind consumed with anxiety about providing for yourself and your family.

Before I had started growing my uMkhiwane tree, while I was still at Ezemvelo, I had always used the resources I had to help my community. In 2009, I'd been approached by Siyabonga Mhlongo and Sanele Hlatswayo, who wanted to use some underutilised land in the community to create a vegetable garden. I knew the first step would be to fence the garden to protect it from wandering wild and domestic animals. At first, we just used branches from thorn trees.

I also encouraged Siyabonga and Sanele to get young people involved in the project, as their chances of success would be much greater with more hands to do the work. They struggled to generate interest at first. Agriculture is not popular among the young adults in our community. Working with the soil is seen as degrading or dirty. This is especially true of the men, because traditionally women worked with the soil. There were some men who said they would be willing to help with building fences, but not with planting, weeding and watering. That was women's work. Even some of the young women were reluctant to get their hands dirty.

I spoke to them: 'What does it matter if your hands are dirty if it means getting food for your family? A leopard has a beautiful coat of spots, but he will not think twice to run through mud to catch a warthog. Do you want your children to go hungry so you can keep your hands clean? Do you want to wait for the government to drop food from the sky?'

We welcomed anyone who was willing to work. Slowly, the numbers grew.

One of the elders in the village offered us some land that he was not using, a maize field that was no longer productive due to the droughts caused by climate change. Water was not a problem, thanks to a project dating back to 1994 to pump water from the river. Each homestead pays about R20, which goes into an account and can be used to fix the pipes or the pump should they break. This project is still running smoothly. All we needed was a big JoJo tank on site, a pump and pipes to connect it to the main water reservoir. Over time, we obtained these things with funds from groups who did trails with me.

We knew that the land was tired, and that the first thing we needed to do once we had fenced it was to feed the soil. We gathered cow dung, chicken manure and composted vegetation, and used hoes to scrape it into the red soil to create a rich, fertile bed. As we worked in that first garden, birds flocked around us, pecking at the insects exposed by our digging; the soil was alive with hundreds of tiny creatures, grateful for the new nutrients we were bringing to them. The dust coated my nostrils and throat; I could taste and smell the earth. Sweat ran down our faces, but our tired, burning muscles were given new energy by our songs that echoed through the hills.

My years in the wilderness have taught me that in nature there is no waste. Everything is reusable and can be recycled – humans are the only creatures to break this sacred cycle. An elephant may destroy many trees in its life, but its dung provides rich fertiliser. Many creatures feed on dung, but termites play a specific role in nourishing the soil. The stickiness of their bodies' excretions help to bind the nutrients from the dung to the soil; bacteria in their gut help to extract nitrogen from the air and fertilise the soil. The partnership between the elephants and termites creates a nourishing base for plants to grow in, while creatures such as dung beetles ensure that undigested seeds in the dung are spread far and wide.

We used these lessons in our gardens, digging in the umquba – old cow dung that has been buried deep under the surface of the cattle kraal – the chicken manure and composted vegetation, which we created by burying cut grass, leaves and kitchen scraps, and watering it. We planted peanuts to help secure nitrates.

The garden gained a boost when the Menzies Aviation group donated funds to help with connecting irrigation pipes and fencing. This was when the first seeds of my uMkhiwane tree were sown, as I could see how the generous contributions from those who had been on wilderness trails could bring so much benefit to my community.

Over the months, we developed a way of working that was sustainable – and I was able to use the learnings of the first vegetable garden to create six more when I was approached by youth in other areas. We received advice from the African Conservation Trust on permaculture and organic farming methods. We managed pests by spraying the crops with dishwashing liquid mixed with water. The groups agreed on the main crops to be sown, but people were also free to plant any vegetables they wanted for their families.

Anyone in the group could harvest whatever they needed, as long as enough was left for the others, and also for sales. We supplied local schools for feeding projects, and sold to community members at a weekly market. Any profits were first ploughed back into the project to buy seeds and equipment, then divided among the group so that they had money for school fees and other necessities.

I started the gardens with a view to feeding the bodies of my community. But I could soon see that this project was feeding far more than that. Following the patterns of nature in the gardens, their hands in the soil, the youngsters brought themselves into deep connection with the earth. Anyone who has worked with the earth will tell how healing it is to touch the soil.

I could see the difference the gardens and all they generated were making to people – their growing confidence, their sense of pride and dignity. I remember hearing one man say, as he knelt between the rows of spinach and beetroot, digging out weeds, 'I was so stressed when I left home, but coming here and seeing that there is new life emerging gives me hope that I can manage my problems.'

I realised that the gardens were not only feeding their bodies, but also their souls, by helping them to forge deep connections to the spiritual energy of the earth. Just as elephant dung needs termites and many other small insects and microbes to become fertile soil, so we need collaboration with other life forms to generate strong bodies, minds and spirits. Creating healthy soil reminds us of the collaborative relationships in nature, the intricate, infinite,

Community gardens feeding souls

entangled processes of giving and taking, the contributions of small creatures to entire ecosystems.

At the time of writing, the gardens have been going for eleven years. They are organic and self-sufficient and generate income. I help with mentoring and guiding, but the young adults run and work the gardens themselves.

Indeed, eating food you have grown yourself feeds many hungers.

EPILOGUE

———•———

IN APRIL 2020, THE SEASON OF FEAR WAS UPON US. MY BROTHER, Siyabonga, was down at the kraal, violently attacking the innocent soil with a hoe, sending earth clods fleeing in alarm. I could hear him cursing the new virus that was shadowing our country. My brave, never-defeated mother began singing softly to calm him and encourage him to measure his pace as he hoed the soil to prepare it for planting. We Africans heal our grief and sorrow through song.

Along the gravel road above my house was a steady stream of the Toyota minibuses that are used as public transport in the rural areas. They were bringing people home from all corners of South Africa, all trying to escape the jaws of this unknown disease. I watched them drive past, wondering whether their passengers were carrying the virus with them, if they would bring it to the vulnerable elders of our community, for how could we 'socially distance' in our small homes? In rural communities, our elders are seen as a vital link to the spirit world because of their ability to communicate with ancestors. Losing them would leave our souls naked and confused.

I knew that most of the people pouring back into our area had lost their jobs, or would lose them in the coming months, piling yet more hardship onto our already embattled community. My own inbox was full of emails cancelling trails, and my livelihood had vanished overnight.

I looked through my WhatsApp messages, seeing more cancellations, then came across one from my friend Bridget, an author from Cape Town. She'd called earlier in the year to offer to help me write a book about my journey. 'Sicelo, you've helped me so much with my own writing. Let me help you get this book out of your head and onto paper.'

The idea of a book had been brewing for some years. But how could I write a book now? I was drowning in fear of the coming months. How could putting words on paper help me?

Still, there was not much else I could do. I collected my journals, and read over my notes. I sat down at my computer, and began to write.

My name is Sicelo Cabangani Mbatha. My wilderness name is Bhubeselimnyama – Black Lion …

The words came slowly at first, a trickle, then a stream, then a flood, like a river swollen by summer rains rushing to the sea. Writing this book brought my beloved wilderness back to me. Lockdown prevented me from walking in the tracks of the elephants in the iMfolozi, or from climbing the high slopes of the uKhahlamba-Drakensberg. Yet, as I wrote, it felt as if I was walking those paths and breathing that mountain air, for even just the memory of them brought a flood of the healing energy. I felt a resurgence of spiritual upliftment, as if I'd been swept up into the thermals and was gliding, gliding, gliding, high above the grassy slopes of the mountains, like the tawny lammergeiers and white-backed vultures.

Writing this book offered me a refuge from my sadness but it also reinvigorated my soul, and gave me the strength to redouble my efforts to help the community, to use our limited resources to build on the gardens we had already created and grow new gardens in each homestead, to be self-reliant and resilient, to grow back from the fires of the virus as the sugarbush grows back after fires on the mountain.

———

In the August before this tiny virus disrupted our world, I was on trail with people from Germany. August is the month of winds, the month of thirst, and also the month of transformation. In the afternoon we dug holes in the dry riverbed to find water, then sat in the shade of the umkhiwane trees on the riverbank and watched as baboons and impalas came down and gratefully drank from the small waterholes we had created. I felt so moved that our simple act of seeking water for ourselves had brought benefit to other creatures. It expressed such a clear vision of how we could live in the world so differently. Instead of building dams to hoard water for ourselves, we could dig waterholes to share with other creatures. Instead trying to keep safe by building walls that isolate us from one another and the natural world, we could build bridges of connection, collaborate in this sacred dance of life.

As we set up for the night, the wind swirled around us, restless, blowing dust and sand into our eyes, singing through the dry grass, spinning leaves into spirals that scudded across the landscape. A lonely baboon shouted out from a nearby tree; from a far ridge, the fading cries of the zebra yearlings were carried on the wind. Having dust blown into your eyes is not a comfortable thing, yet it seemed that this wind was warning us, or perhaps inviting us. 'Vukani!' it was calling. 'Wake up! You humans have been sleep-walking for too long, stumbling through life, blocking your eyes and ears to the sorrows of the world. Wake up!'

The storm broke as dark cumulonimbus clouds roiled on the

horizon, slashed with bright flashes of lightning. As we ate our simple meal in silence, soft rain fell on the grateful soil. A Burchell's coucal sang its praise song for the rain, its mellow warbling harmonising with the quietening rumbles of thunder. Later, we were lulled to sleep by the patter of drops on the flysheets and the high trills of crickets.

I woke early and sat watching the age-old ritual of the dawn as the sun slowly rose behind the umbrella thorn trees. The river had come to life again, a stream of clear water running over the golden sands. Millipedes and other insects emerged from their winter slumbers, invigorated by the life-giving rain. If the wind was a warning, the rain and the dawn that followed were a promise, a promise of a re-enlivened world that may come if we wake up to its possibilities.

The glory of the morning breeze, the birds singing under the piercing, blue sky, the white mist smoking above the iMfolozi's valleys, became my prayer for the world, a prayer that we may all dance in peace, a prayer for healing, for justice and equality, for the true prosperity that comes with unconditional – and non-transactional – love.

We humans have acquired much information, but have lost much wisdom. We have enough information to dig out the last lump of coal, to drain the last drop of oil, yet we lack the wisdom to protect our one and only home and source of life. The guinea-fowl is often mocked for being stupid. Yet it has the wisdom to hide its chicks from the hawk – while we fail to protect our children from violence, hunger, environmental catastrophe.

I know that nature is where we may rediscover our lost wisdom. I have learnt about the power of forgiveness from the jaws of a crocodile. I have learnt about grief from a bereaved baboon. I have learnt about resilience from the burnt protea. I have learned about ubuntu from the elephant. By walking on ancient elephant pathways, I have learnt that love can wash the dust from our eyes,

Everything I need I carry on my back

and I have seen what wisdom these pathways have brought to so many others. The crystal-clear waters of the iMfolozi rivers reclaim me, you and us. Their waters bring us the clarity we need to see ourselves for who we really are, now, and who we could be.

This book is an invitation, to step into the wilderness with me, to heed the warning of the wind and wake up to the vibrancy of the earth, to open yourself to the transformation we need to heal our world.

I hope that you will join me on trail one day, that we will hoist our packs onto our shoulders, smile up at the wide blue sky, set our feet on the path, and walk together.

Past the bushwillows and camphor trees, into a better world.

ACKNOWLEDGEMENTS

BLACK LION HAS BEEN A VERY SPECIAL PROJECT TO ME, ONE THAT HAS invited me to visit my vulnerability, my fears and also my hopes. Like an African child who is raised by a village, this story has been raised by many hands. I am deeply grateful to my family (my mom Siphiwe Mbatha, brother Siyabonga, sister Makhosi and nephew Siboniso Msweli), and to all those whose enthusiasm and energy transformed my vision of this book into reality, especially Bridget Pitt – your commitment and dedication move me.

I would like to thank Jeremy Boraine, for believing in this book and making it possible, and Caren van Houwelingen, Angela Voges and the rest of the team at Jonathan Ball for all their support and hard work in making it the best it could be.

I express my special thanks to Bruce Dell, Cheryl Curry, Ian Read, Paul Cryer, Mandla Buthelezi and Mandla Gumede for their endless support and love

To Richard Mchunu, Sabelo Msweli, Nkalakatha Nxumalo, Baba Thabethe, Baba Gumede, Dumisane Khumalo, Sonto Mthembu,

Dave Robertson, Craig Reed, Kim Gillings and Lawrence Monroe, I deeply value your guidance.

To all the communities and wilderness guides who attended my seminars in Germany and Austria, you give me hope that my work is reaching many hearts.

To Richard Knight, Geseko von Lüpke, Wolfgang Peham, Barbara Deubzer, Peter Jentzen, Christa Regenfuss, Claudia and Phillip Hoerl, Markus Hirschmann, Thomas Weber, Joscha Grolms, Elisabeth Demeter, Alexander and Megan Kelly, and Steven Vervloet, thank you for bringing light during the dark times of my life (ubuntu).

To Dudu, my wife and best friend, to my daughter Ntando for your presence, and my sons Andiswa and Asande, you are all the sunlight that lights my path.

To my communities, Ezigwilini, Hlambanyathi and Emafamu, thank you for raising me to become what I am today. I love you all.

SPECIES LIST

———•———

English Name	Scientific Name	IsiZulu Name
Aardvark	*Orycteropus afer*	iSambane
African elephant	*Loxodonta africana*	iNdlovu
African fish eagle	*Haliaeetus vocifer*	iNkwazi
African paradise flycatcher	*Terpsiphone viridis*	iNzwece
African scops owl	*Otus senegalensis*	iSkhova
African wild dog	*Lycaon pictus*	iNkentshane
African wood owl	*Strix woodfordii*	iNkovane, uMabhengwana
Bateleur	*Terathopius ecaudatus*	iNqunqulu
Bee-eaters	Meropidae	Isdlanyosi (differs between villages)
Black-backed jackal	*Lupulella mesomelas*	iKhanka
Black-bellied bustard	*Lissotis melanogaster*	iNgagalo
Black mamba	*Dendroaspis polylepis*	iMamba emnyama
Black monkey thorn	*Acacia burkei* (now *Senegalia burkei*)	uMkhaya

Black rhinoceros	*Diceros bicornis*	uBhejane
Blacksmith lapwing	*Vanellus armatus*	iNdudumela
Black wildebeest	*Connochaetes gnou*	iNkonkoni
Brown scrub robin	*Cercotrichas signata*	uShesheshe
Buffalo thorn	*Ziziphus mucronata*	uMlahlankosi
Burchell's coucal	*Centropus burchellii*	uFukwe
Burchell's zebra	*Equus quagga burchellii*	iDube
Bush babies	Galagidae	iSinkwe
Bushbuck	*Tragelaphus scriptus*	iMbabala
Bushman plum	*Acokanthera oppositifolia*	iNhlungunyembe
Bush medlar	*Vangueria infausta*	uMviyo
Bushwillow	*Combretum erythrophylum*	uMbondwe
Cabbage tree	*Cussonia spicata*	uMsenge
Cape buffalo	*Syncerus caffer caffer*	iNyathi
Cape glossy starling	*Lamprotornis nitens*	iKhwezi
Cape porcupine	*Hystrix africaeaustralis*	iNgungumbane
Cape turtle dove	*Streptopelia capicola*	iHope
Chacma baboon	*Papio ursinus*	iMfene
Cloud cisticola	*Cisticola tetrix*	uNcede
Common crowberry	*Rhus pentheri*	iNhlokoshiyane
Common duiker	*Sylvicapra grimmia*	Impunzi
Crested francolin	*Dendroperdix sephaena*	iSikhwehle
Crested guineafowl	*Guttera pucherani*	iMpangele
Dassie (rock hyrax)	*Procavia capensis*	iMbila
Dung beetles	Scarabaeinae	iBhungane
Eland	*Taurotragus oryx*	iMpofu
Emerald-spotted wood dove	*Turtur chalcospilos*	iNkwambazane
Fiery-necked nightjar	*Caprimulgus pectoralis*	uZavolo
Giant kingfisher	*Megaceryle maxima*	isiVuba
Giraffe	*Giraffa*	iNdlulamithi

Species list

Golden orb spider	*Nephila*	Isicabucabu
Goliath heron	*Ardea goliath*	uMazazingwenya
Gorgeous bushshrike	*Telophorus viridis*	iNgulube
Great egret	*Ardea alba*	iLanda
Greater honeyguide	*Indicator indicator*	iNhlava
Greater kudu	*Tragelaphus strepsiceros*	uMgakla
Hadeda ibis	*Bostrychia hagedash*	iNkankane
Hamerkop	*Scopus umbretta*	uThekwane
Hippopotamus	*Hippopotamus amphibius*	iMbom, iMvubu
Hyena (spotted)	*Crocuta crocuta*	iMpisi
Impala	*Aepyceros melampus*	iMpala
Jackal buzzard	*Buteo rufofuscus*	iSkhobotho
Kei apple	*Dovyalis caffra*	uMqokolo
Lammergeier (bearded vulture)	*Gypaetus barbatus*	uKhozi Lwentshebe
Lappet-faced vulture	*Torgos tracheliotos*	Buka izwangomoya
Large-leaved rock fig	*Ficus abutilifolia*	uMkhiwane
Leopard	*Panthera pardus*	iNgwe
Lilac-breasted roller	*Coracias caudatus*	iFefe
Lion	*Panthera leo*	iBhubesi
Magic guarri bush	*Euclea divinorum*	iDungamizi
Marula	*Sclerocarya birrea*	uMganu
Mountain reedbuck	*Redunca fulvorufula*	iNxala
Mountain stinging nettle	*Obetia tenax*	uLuzi
Nile crocodile	*Crocodylus niloticus*	iNgwenya
Nyala	*Tragelaphus angasii*	iNyala
Oldwood	*Leucosidea sericea*	uMtshitshi
Ox-eye daisy	*Leucanthemum vulgare*	iMpila
Phragmites reed (common reed)	*Phragmites australis*	uMhlanga
Pied kingfisher	*Ceryle rudis*	iHlabahlabane
Red-billed oxpecker	*Buphagus erythrorhynchus*	aMahlalanyathi

Red-chested cuckoo	*Cuculus solitarius*	uPhezukomkhono
Red grass	*Themada triandra*	uMfazibomvu
Red ivorywood	*Berchemia zeyheri*	uMncaka
Rock monitor	*Varanus albigularis*	iMbulu
Sausage tree	*Kigelia africana*	uMvongothi
Scarlet-chested sunbird	*Chalcomitra senegalensis*	iNcwicwi
Scented-pod acacia	*Acacia nilotica* (now *Vachellia nilotica*)	uMnqawe
Shepherd tree	*Boscia albitrunca*	uMsenge
Short-clawed lark	*Certhilauda chuana*	uNqangendlela, uQwaqwashe, uNonqashi
Sickle bush	*Dichrostachys cinerea*	uGagane
Southern masked weaver	*Ploceus velatus*	iHlokohloko
Southern red bishop	*Euplectes orix*	iBomvana
Speckled pigeon	*Columba guinea*	iJuba
Splendid thorn	*Acacia robusta (now Vachellia robusta)*	uMngamanzi
Spotted eagle-owl	*Bubo africanus*	iSkhova esikhulu
Spotted thick-knee	*Burhinus capensis*	uMbangaqhwa
Spur-winged goose	*Plectropterus gambensis*	iHoye
Sugarbush	*Protea caffra*	Isqalaba
Sweet thorn	*Acacia karroo* (now *Vachellia karroo*)	umuNga
Sycamore fig	*Ficus sycomorus*	uMkhiwane
Tamboti	*Spirostachys africana*	uMthombothi
Torchwood	*Balanites maughamii*	uMnunu, uGobandlovu
Trumpeter hornbill	*Bycanistes bucinator*	uMkholwane
Turpentine grass	*Cymbopogon*	iMbubu, iMbanjane
Umbrella thorn	*Acacia tortilis* (now *Vachellia tortilis*)	uMsasane
Verreaux's eagle	*Aquila verreauxii*	uKhozi

Species list

Verreaux's eagle-owl	*Bubo lacteus*	iFubesi
Waterbuck	*Kobus ellipsiprymnus*	iPhiva
Weeping boer bean	*Schotia brachypetala*	uMgxamu
Western cattle egret	*Bubulcus ibis*	iLanda
White-backed vulture	*Gyps africanus*	iNqe
White rhinoceros	*Ceratotherium simum*	uMkhombe
Wild basil	*Clinopodium vulgare*	uMnandi
Wild camphor bush	*Tarchonthus camphoratus*	iQgeba
Wild potato bush	*Coleus esculentus*	uMhlaza
Wild syringa	*Burkea africana*	uMnondo
Woolly caper bush	*Capparis tomentosa*	iQwaningi
Yellow-billed kite	*Milvus aegyptius*	uNhloyiya

2